Better Homes and Gardens®

WOODWORKING PROJECTS
YOU CAN BUILD

BETTER HOMES AND GARDENS® BOOKS

Editor in Chief: James A. Autry
Editorial Director: Neil Kuehnl
Executive Art Director: William J. Yates

Editor: Gerald M. Knox
Art Director: Ernest Shelton
Associate Art Directors: Neoma Alt West,
    Randall Yontz
Assistant Art Director: Harijs Priekulis
Copy and Production Editors:
    David Kirchner, Lamont Olson,
    David A. Walsh
Senior Graphic Designer: Faith Berven
Graphic Designers: Linda Ford,
    Sheryl Veenschoten, Thomas Wegner

Building and Remodeling Editor: Noel Seney
Building Books Editor: Larry Clayton

Woodworking Projects You Can Build
Editors: Noel Seney, Larry Clayton
Copy and Production Editor: David A. Walsh
Graphic Designer: Thomas Wegner
Contributing Writer: Gary Boling
Exploded Drawings: William C. Schuster

Project Design Credits
Stacy Strand, 6, 26; Dave Ashe, 9, 20,
22, 24, 25, 27, 31, 37, 53, 62, 68, 80;
Jerry Ross and Debbie Schmitz 13; The
Design Post, 14; Stephen Mead, 15, 44,
82; Richard Banks, Architect, 16; Camille
Lehman, 19; Herman Cortes, 21, 34;
Color Design Art, 29; Peggy Walker, 30;
Gary Gerber, 33; Anita and Neal
Orenstein, 35; Suzy Taylor, ASID, 42, 49,
73; James P. Abeloe, 45; Chet Ross, 47;
Ernst O. Meissner, AIA, 50; Donald K.
Olsen, AIA, 52; Garth Graves, 55, 74;
Roland L. Molen, 57; Rod V. Luccioni,
Architect, 58; The Just Plain Smith
Company, 61; G. E. Sparrow, 63; Kitchen
Designs, Inc., 64; Kenneth H. Peterson,
67; Shirley Marks–Roger De Vito Assoc.,
70; Denise Hodges, 79.

CONTENTS

HALF-DAY PROJECTS

If you think that attractive projects take forever to build, we've got some good news — and some good ideas — for you. In this opening chapter, we've assembled a collection of projects that can be knocked together in an afternoon or less.

Since almost everyone needs more storage space, we've included lots of projects for stashing everything from stereos to socks to plants to pickles. In addition, you'll find seating and lighting ideas, a room divider, and accessories that are sure to brighten a dreary corner.

These projects not only are quick and easy to build, but also are easy on the budget. In fact, some of the units featured require few materials other than those you may be able to find in your scrap lumber bin.

The next time you have a little time on your hands, give one of the projects in this chapter a try. You'll be pleasantly surprised by the handsome results.

SPACE-SAVER STEREO CENTER

Finding adequate storage space for stereo components can be a very big problem. And that's exactly why this unit is just right for so many situations. Its upright, wall-hugging design requires very little floor space, yet the shelves can comfortably and safely accommodate most stereo equipment.

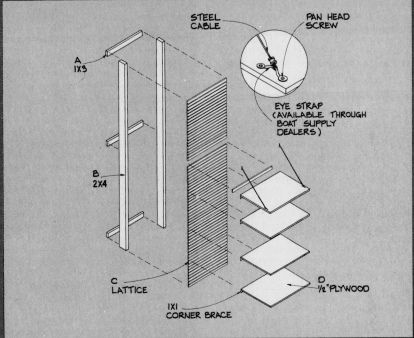

STEEL CABLE
PAN HEAD SCREW
A 1x3
EYE STRAP (AVAILABLE THROUGH BOAT SUPPLY DEALERS)
B 2x4
C LATTICE
1x1 CORNER BRACE
D ½" PLYWOOD

1 The unit shown features 23 × 18-inch shelves. If these are too big or too small for your equipment, adjust the dimensions accordingly.

2 From 1 × 3 lumber, cut three stretchers (A). Cut two 2 × 4 uprights (B) just shy of floor-to-ceiling length. Working on a flat surface, sandwich the 1 × 3s between the uprights, and join with glue and nails. Secure the unit to the wall studs.

3 Cut lengths of lattice (C) to extend across the uprights. Plan where you want to place the shelves, then nail lattice across uprights. For now, skip a strip close to each shelf to facilitate running wiring between your stereo components. Finish the wood as desired.

4 Cut the needed number of shelves (D) from ½-inch plywood. Cover exposed edges with wood veneer tape, or if you're painting the shelves, fill voids with wood putty. Finish as desired. Turn the shelves upside down and screw corner braces where shown.

5 Position and secure the shelves as shown.

6 After running the wiring to various components, nail the remaining lattice strips into position.

Materials: ½-inch plywood, lattice, 2 × 4 and 1 × 3 lumber, 1 × 1-inch corner braces, eye straps, ⅛-inch steel cable, crimp bands, wood veneer tape and contact cement or wood putty, and desired finish.

NATURAL-LOOK PLANT CENTER

Even alone, a small houseplant does wonderful things for its surroundings. But when grouped together, houseplants are even more impressive. This notch-together unit will corral your plant crowd in a hurry.

1 This plant stand measures 30 inches square. The lowest shelf is about 13 inches high.

2 Begin by cutting seven like-sized pieces of 1 × 8 lumber for the sides and the ends of the unit (A,B,C).

3 Cut ¾ × ¾-inch-deep notches in the 1 × 8s as shown in the sketch. Note the absence of notches at the bottom of members A and at the top front of members C. Also cut ¾ × ¾-inch-deep rabbets in two of the B members (see sketch).

4 Cut a length of 1 × 6 for the top shelf (D). This member is the same length as members A, B, and C. Then cut a ¾ × ⅜-inch groove down the center of the board.

5 Cut five more lengths of 1 × 6 lumber for the lower shelves (E). Make these members 1½ inches shorter than the top shelf (D). Then cut the 1 × 1 ledgers (F,G) to size.

6 Glue and nail the ledgers to the inside of side members C as shown in the sketch.

7 Assemble the plant stand, using glue and nails to secure the joints.

8 Finish the unit as desired.

Materials: 1 × 8, 1 × 6, and 1 × 1 lumber; glue; finish nails; and choice of finish.

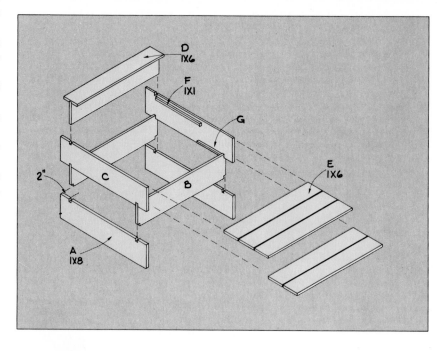

IN-A-JIFFY
COFFEE TABLE

This clean-lined unit must surely rank as one of the easiest-ever projects to build. But that doesn't stop it from being an interesting and functional piece of furniture. The tabletop is a ½ × 16 × 54 piece of plate glass.

1 The table base shown measures 48 inches long and is about 14 inches high, including the casters and the ½-inch-thick plate glass top. But feel free to make any size adjustments necessary to suit your situation.

2 Using good-quality 2 × 12 lumber, cut two lengths for the top and bottom (A), and two lengths for the uprights (B). Miter the ends of each piece. (If you are unable to do this with the equipment you have, have the lumberyard cut the lumber for you.) Or shorten the uprights, lengthen the top and bottom members, and use simple butt-joint construction.

3 Construct the unit, using finishing nails and glue. Finish the table as desired.

4 Screw four 2-inch plate casters to the bottom of the unit. Be sure to position them back in from the edges. Top the table with a piece of ½-inch-thick glass (C). The weight of the glass should keep it in place. You may want to put circular rubber pads between the table and glass.

Materials: 2 × 12 lumber, ½-inch-thick plate glass with seamed edges, four 2-inch plate casters, glue, finish nails, and your choice of finish.

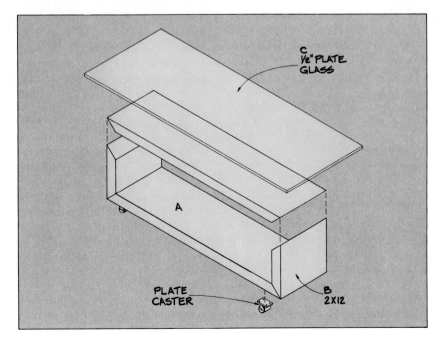

QUICK-CHANGE SHELVING SYSTEM

This slip-together unit makes displaying your collectibles a creative outlet. When you tire of one arrangement, move the shelves around for a new look. Made from sturdy ¾-inch plywood, the shelves stand up quite well to heavy loads.

1 Though this unit extends almost from floor to ceiling, don't hesitate to alter the dimensions to suit your needs.

2 From ¾-inch plywood (red oak plywood was used here), cut three 7½-inch-wide pieces to serve as the backs of the uprights (A).

3 Cut six 6½-inch-wide shelf supports (B). Before cutting notches, check floor level to make sure shelves will be level. If floor is not level, plan where bottom notch will fall, and level across all three uprights at this point. Make all measurements for notches relative to this mark.

4 Using a saber or crosscut saw, cut ¾ × 5-inch notches in the desired positions. Glue and nail the shelf supports to the plywood backs (A). Sink the nailheads. If desired, add metal corner braces to the inside corners of the uprights for strength.

5 For the shelves, cut 10-inch-wide pieces of ¾-inch plywood (C,D,E) to the lengths desired.

6 Cover exposed plywood edges with wood veneer tape, or, if painting, fill voids with wood putty. Fill nail holes and finish as desired.

7 Fasten the uprights to the wall studs at 32-inch intervals using screws. Counterbore the screwheads and insert dowel pins to conceal the voids. Insert shelves into their slots.

Materials: ¾-inch plywood, metal corner braces (optional), wood putty, veneer tape and contact cement (optional), screws, wood dowel, and finish.

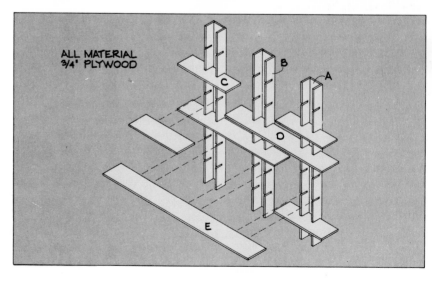

ALL MATERIAL ¾" PLYWOOD

WINDOW PLANT GARDEN

If your kitchen windowsill is lined with small jars filled with plant cuttings, here's the perfect project to create order out of chaos. Collect several jars with similar diameters, a crop of promising plant clippings, and you're on your way to a bounty of healthy new plants.

1 This window unit measures 30 × 37½ inches. If necessary, adjust the dimensions so the unit's uprights will rest squarely on the window's trim pieces.

2 Cut uprights (A) from 1 × 6s. Round the corners to a 4-inch radius with a saber saw.

3 Cut three shelves (B) from 1 × 6s. Then cut a series of evenly spaced holes in the shelves for the jars.

4 Mark the shelf positions on the uprights. Measure the "drop" of the jars, and mark the position of the ½-inch-diameter dowel supports (C). Clamp the uprights together and drill holes for the dowels. Using butt joints, glue and nail the shelves to the uprights. Sink the nailheads and fill the holes. Glue dowels into place.

5 Sand the unit smooth. Finish as desired.

6 To hang the unit, drive screw eyes into the window's top trim. Suspend chain from the screw eyes. Secure two screw eyes to the uprights so the unit will hang straight. Hook the bottom of the chains into the screw eyes.

Materials: 1 × 6 lumber, ½-inch dowel, glue, nails, screw eyes and S-hooks, chain, wood putty, and finish.

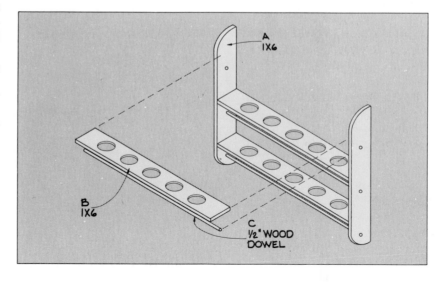

OVERHEAD PLATFORM STORAGE

If you're like most people, you've never thought of a hallway when looking for more storage space. But this project may just change your way of thinking. You'll be amazed at how much platforms like these will hold, yet they're conveniently out of the way. They are ideal for storing seldom-used and seasonal items such as luggage and blankets.

1 Plan the storage units to be as wide as the hallway. Place them level with the top of the doors; make them short enough so you can reach the items easily.
2 Nail 1 × 2 ledgers (A) to the wall studs along the hallway wall. Cut a piece of plywood (B) to span hallway. Glue and nail it to the underside of the ledgers.
3 Cut 1 × 3 end pieces (C) for retainer lips; glue and nail them to edges of plywood.
4 Sand the unit smooth and finish to coordinate with wall paint or woodwork color.

Materials: ¾-inch plywood, 1 × 2 and 1 × 3 lumber, glue, finish nails, and choice of finish.

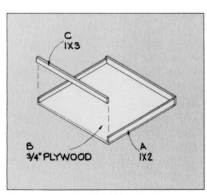

OVER-DOOR STASH-ALL

Here's another clever variation of the overhead storage idea. You can scatter several of these around the house in a matter of hours. Either emphasize the unit with trim paint or camouflage it by matching the ceiling color.

1 This unit is as wide as the door and about 60 inches long. Adjust the dimensions.

2 Cut two 1 × 2 ledgers (A). Then cut two ¾-inch plywood pieces (B,C). Glue and nail the plywood into an L-shape. Fortify this joint with metal corner braces.

3 Glue and nail the 1 × 2 ledgers to the plywood L as shown. Then with a helper, lift the unit into place and fasten it to wall studs and ceiling using screws. If this isn't possible, anchor the unit with toggle bolts.

4 Cut two pieces each of 1 × 3 trim (D,E), miter the corners as shown, and glue and nail to plywood edges.

5 Sand and finish the unit.

Materials: ¾-inch plywood, 1×3 and 1×2 lumber, corner braces, glue, nails, and desired finish.

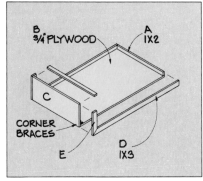

DROP-LEAF COFFEE TABLE

Drop-leaf tables are furniture favorites. And little wonder, since their functional design provides lots more table space when the leaves are raised. We've updated this style and created a classy-looking coffee table for you to enjoy.

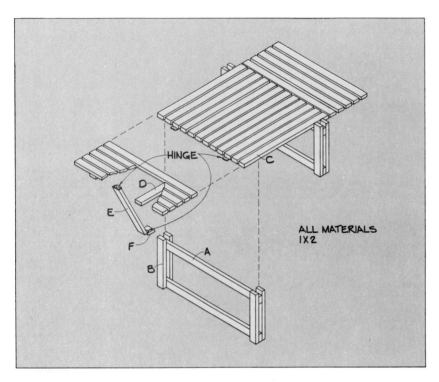

HINGE

D

E

C

F

A

B

ALL MATERIALS
1 X 2

1 The unit shown is 25⅛ × 40 inches when the leaves are upright. It measures 11 inches high.
2 From 1 × 2 lumber, cut four base crossmembers (A), and eight uprights (B). Construct each support by sandwiching the crossmembers between the uprights. Nail and glue together.
3 Again using 1 × 2s, cut the tabletop members (C). Then with the bases perpendicular to the floor, glue and nail top 1 × 2s to the support crossmembers so the tabletop overlaps the supports by 1½ inches on each end. Space the members evenly. To facilitate finishing, you may wish to sand and finish the top pieces before securing them.
4 Cut four 9-inch-long crossmembers (D) for the drop leaves. Cut ten 1 × 2 top pieces (same size as C) for the leaf surfaces. Glue and nail the 1 × 2s to the crossmembers; space the members evenly. Hinge the drop leaves to the ends of the table.
5 Invert the unit so the legs are pointing up. Measure the distance between the edge of the drop leaf and the lower crossmember of the base assembly. Cut leaf braces (E,F) so they'll hold the leaves level when in their erect position. Angle-cut the braces as necessary for a snug fit. Attach the braces to each other and the underside of the table with butt hinges.
6 Fill all nail holes; sand smooth and finish as desired.

Materials: 1 × 2 lumber, butt hinges, glue, finish nails, wood putty, and desired finish.

PORTABLE
PLANT
SHOWCASE

At home either indoors or out, this roll-anywhere plant stand will provide center-stage treatment for your showy houseplants. If used indoors, be sure to line the inside of the container with plastic to prevent mishaps while watering the plants.

1 The planter shown measures 18 inches square and 24 inches high. Alter the dimensions to fit your space and plants.

2 Cut the planter's sides (A) from ½-inch rough-sawn cedar plywood. Miter the edges of each piece. Cut four 1 × 1 ledgers (B) to size, then glue and nail them to the inside of the planter's sides as shown (position two at the height you want the plant platform). Assemble the planter, using nails and waterproof glue.

3 Cut the top platform and a base (C) from rough-sawn cedar plywood. Glue and nail them to the 1 × 1 ledgers. (If you'll be using the planter for potted plants, cut circles in the top platform to hold the pots.)

4 To trim the planter, cut members D, E, and F to size; miter the appropriate ends (see sketch); and glue and nail the trim pieces to the sides of the planter.

5 Fill plywood edges and apply a water-resistant finish.

6 Fasten plate casters to the underside of the base. See page 92 for installation how-to.

Materials: ½-inch exterior cedar plywood, 1 × 1 and 2 × 2 lumber, finish nails, waterproof glue, plate casters, and finish.

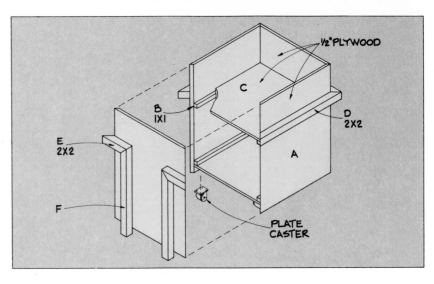

SPACE-SAVING STACKING TABLES

In small rooms especially, floor space is always at a premium. Here's a practical table project that caters to many limited-space situations: Three tables in the space of one!

1 The largest table of this trio measures 18 × 19 inches, and is 14 inches tall. You can alter these dimensions somewhat, but keep the height under 16 inches so the tables won't be top-heavy.

2 Cut the top and sides of the largest table (A,B). Glue and nail the members together as shown in the sketch.

3 Measure the inside dimensions of the table. Allowing ⅛-inch all around for clearance, cut the pieces for the next-smaller table (C,D). Assemble the table as before.

4 Measure the inside dimensions of the second table and cut smallest table pieces (E,F). Allow ⅛ inch for clearance. Assemble the table as before.

5 Fill exposed plywood edge voids with wood putty. Sand the edges smooth; paint the tables. Let dry completely.

6 To duplicate the center stripe on the tables, mark the outline of an 8-inch-wide stripe down the center of each table, then mask off the remaining area. Paint with contrasting color. If center stripes are to be spray-painted, carefully cover *all* nearby surfaces before spraying. Remove the masking tape as soon as possible.

Materials: ¾-inch plywood, glue, nails, wood putty, and paint.

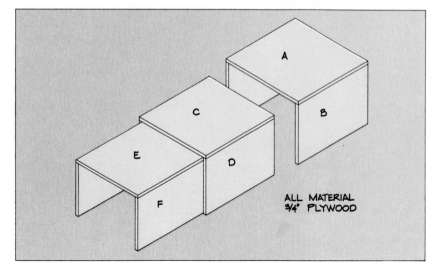

ALL MATERIAL ¾" PLYWOOD

SLEEK SHELF STORAGE

This smooth-looking shelf unit is designed to hold any-thing—from stereo equipment to sculpture to potted plants. Adjustable metal support tubes with shelf standards se-cured out of sight to the rear hold shelves steady. Adjust the length of the shelves and this unit can go anywhere.

1 The unit shown extends from floor to ceiling and is 72 inches wide. If necessary, change the shelf dimensions to fit the space that you have.

2 From 1 × 10 lumber, cut five shelves (A) to the same length. Then cut the desired number of 1 × 10s (B) to serve as vertical dividers. Assemble the com-partmentalized top shelf as shown, using glue and nails. For the three shelves below the box-type top shelf, cut 1 × 10 end pieces (C) and assemble as shown using glue and nails. (Note: If you plan to place the shelf unit away from a corner, add end pieces to the other end of the shelves, too.)

3 For the bottom shelf, cut one piece of ½-inch plywood (D) to size, and notch the front edge to fit around the support tubes. Cut lengths of 1 × 2 (E through K) to size, then glue and nail mem-bers together as shown. Top the 1 × 2 framework with the ply-wood; glue and nail in place.

4 Fill nail holes and exposed plywood edges, sand the unit smooth, and finish as desired. Secure metal shelf standards to the rear side of the metal support tubes. Position tubes and install adjustable shelf brackets. Set the shelves atop the brackets. Secure the top and bottom shelves to the wall studs with metal corner braces.

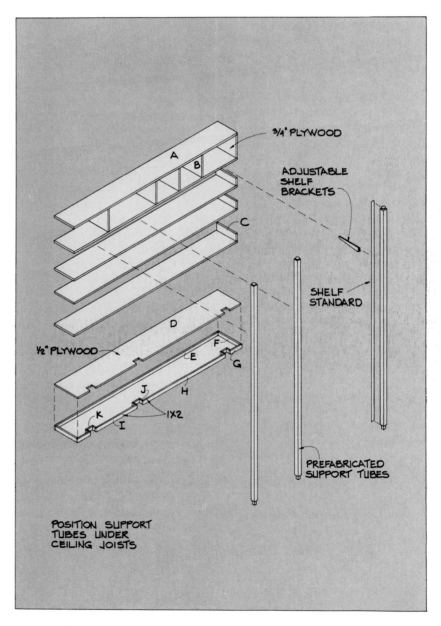

Materials: ½-inch plywood, 1 × 10 and 1 × 2 lumber, glue, metal support tubes, nails, shelf stan-dards and brackets, metal corner braces, wood putty, and finish.

CROSSBUCK-STYLE STUDY DESK

This project is a pleasant blend of the old and the new. Classic crossbuck styling provides plenty of support underneath, and the tempered plate glass top gives the desk a contemporary feel. It goes together rather quickly, too.

1 The table shown measures 36 × 60 × 29½ inches tall, but you can adapt the design to many sizes. Shorten the uprights to create a coffee table, or shorten both the uprights and the crossmembers for an occasional or end table.

2 Begin by cutting notches in the 2 × 4 crossmembers (A). To do this, position them side by side on edge and clamp them together. After determining the lengthwise center of the 2 × 4s, cut 1½-inch-wide notches at a 45-degree angle halfway through them (see notching detail).

3 Assemble the crossmembers, using glue and nails. Let dry. Then cut the crossmembers to length. Those shown are 60 inches.

4 Cut four 2 × 6 uprights (B) to the height desired. Attach the uprights to the crossmembers, using glue and 3½-inch-long wood screws. Predrill the members as shown in the detail to make way for the screws. Fill recesses with dowel pins.

5 Attach furniture glides to the crossmembers.

6 Sand the entire unit smooth. Finish as desired.

7 Attach felt pads to the ends of the crossmembers to prevent scratches on the glass top (C). Position the top.

Materials: 2 × 6 and 2 × 4 lumber (the 2 × 4s should be longer than the finished length of the crossmembers), furniture glides, glue, wood screws, finish nails, ½-inch plate glass with seamed edges, felt pads, wood dowel, and the desired finish.

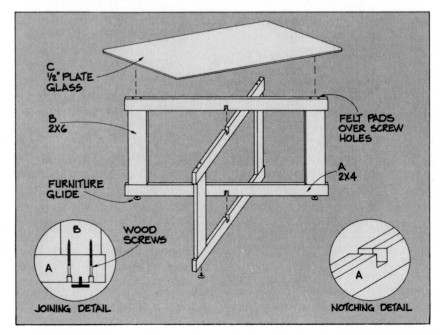

C
½" PLATE GLASS

B 2X6

FELT PADS OVER SCREW HOLES

A 2X4

FURNITURE GLIDE

B
A
WOOD SCREWS

JOINING DETAIL

A
NOTCHING DETAIL

EASY-BUILD KNOCKDOWN COFFEE TABLE

Another crossbuck table that goes together even more quickly than the desk at left is this table. Just make the ta-ble's base as described below from a hollow-core door, then top the table with a 36 × 36-inch surface of your choice.

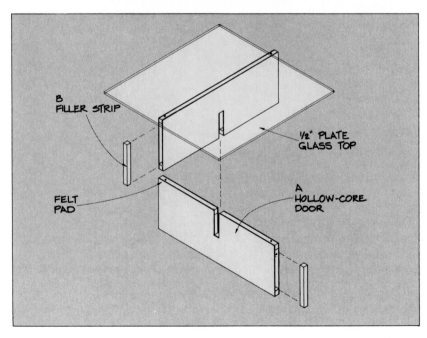

B FILLER STRIP

½" PLATE GLASS TOP

FELT PAD

A HOLLOW-CORE DOOR

1 A 15-inch bifold door serves as the base for the project shown. Select another size door for a taller table. If desired, sub-stitute two thicknesses of ¾-inch interior plywood laminated to-gether for the door. For a natural wood base, buy an oak door, oak veneer tape, and the finish de-sired. If the base will be covered or painted, buy a less expensive door, and your choice of paint or covering.

2 Cut the door (A) in half lengthwise with a fine-tooth handsaw or a saber saw. To pro-tect the veneer, you may want to lay down masking tape before marking the notches' outline. Cut two filler strips (B) to plug the cut edges of the door pieces. Glue into place. If door is to have a natural finish, mask the strips with matching veneer tape.

3 To cut the notches in the base members, first measure the lengthwise center of each. Then carefully mark the outline of each notch. The notches will be as wide as the door is thick and as long as half the width of the members. Cut the notches with a fine-tooth saw.

4 Finish the base pieces as de-sired. Then assemble the base as shown. Fasten a furniture leveler to each leg.

5 Glue small felt pads to the top of the base members, then posi-tion the glass top.

Materials: Hollow-core bifold door, filler strips, ½-inch-thick plate glass, wood veneer tape (optional), furniture levelers, glue, felt pads, and the desired finish.

LATTICEWORK ROOM DIVIDER

Looking for something to create a more interesting entryway in your house? This unit can do just that. And it serves the added function of affording those inside a little privacy from doorway commotion.

1 The room divider shown measures 36 inches wide and extends from floor to ceiling. The lattice inset, which here is 60 inches long, may be lengthened or shortened to suit your needs.

2 From 1 × 4 lumber, cut two uprights (A) to extend from floor to ceiling. Cut two crossmembers (B) to the width desired. Fasten one upright to the wall (you'll have to remove a section of base and any other molding for a flush fit) and the top crossmember to the ceiling. Nail the other upright to the end of the top crossmember as shown. Position the remaining crossmember at the bottom between the uprights, and nail into place.

3 Cut two 1 × 3 lattice supports (C), and cut sufficient lengths of lattice (D) to span the width of the supports.

4 Sand and finish all members.

5 After the finish dries, glue and nail the lattice strips to the supports (C). Note that the strips are offset slightly so they create a privacy-fence effect.

6 Nail the screen assembly to the uprights at the desired height. Add cleats, if desired, for support.

Materials: 1 × 4 and 1 × 3 lumber, lattice, finish nails, glue, and the desired finish.

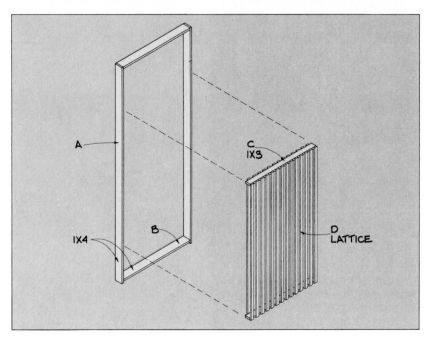

SO-EASY COFFEE TABLES

Affordable, practical, good-looking, and exceptionally easy to build: this striking coffee table duo has it all!

1 These tables measure 21 × 24 × 18 inches high. You can extend the length of the tabletop (you'll need thicker plate glass if you do), but we don't recommend raising the height.

2 For each table cut four 2 × 2 uprights (A) and four 2 × 2 crossmembers (B). Begin by constructing two identical leg assemblies. Clamp one of the crossmembers flush with the top and the outside edge of the uprights. Clamp the other crossmember below the first, leaving ⅜-inch space to allow for the glass top. Drill ¼-inch-diameter holes as shown in the sketch to accommodate the ¼ × 3¾-inch-long bolts that will fasten the members together.

3 With the members still clamped, drill ¼-inch holes down through the crossmembers as shown. Note that these holes are positioned so they don't interfere with those drilled earlier.

4 Sand the uprights and crossmembers, then apply a finish.

5 Bolt the 2 × 2s together. Don't tighten the bolts yet. Then slide a piece of ⅜-inch-thick plate glass (C) between the crossmembers so its ends are flush with the edge of the uprights. Tighten bolts.

Materials: 2 × 2 lumber, ⅜-inch-thick plate glass; ¼ × 3¾-inch machine bolts, nuts, and washers; and the desired finish.

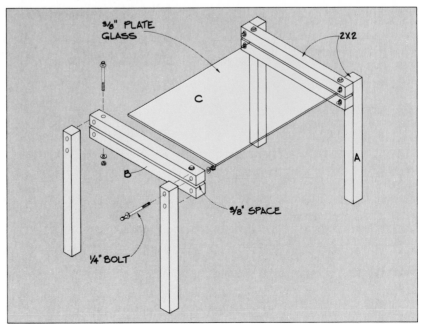

GRILLWORK END TABLE

Here, an iron register grille takes on a new function as the top of this unusual table. Your best bet for finding one of these beauties is a second-hand or junk store. Local contractors involved in remodelings also are good sources.

1 This table measures 17 × 21 × 17 inches high. But you'll have to tailor yours to the size grille you can locate. Keep the height about the same as this one, though. Otherwise, it could become top-heavy.

2 Measure the dimensions of your grille, then cut the sides (A) and the ends (B) one inch longer and wider than the grille. Miter the edges of each piece at 45 degrees for a finished look.

3 In the top edge of the base pieces, cut a ¼-inch-wide rabbet that's as deep as or a little deeper than the thickness of the grillwork.

4 Assemble the base members using glue and finishing nails. Clamp, if possible, until the glue has time to dry.

5 Glue matching wood veneer tape to the exposed plywood edges. Fill all nail holes. Finish the base as desired. Set grille into place, resting it on the rabbeted edge of the base members.

Note: If you can't find a suitable piece of grillwork, a variety of other materials—stained glass, parquet flooring, or ceramic tiles—will work just as well. They'll need reinforcement beneath, of course. If you have found a circular, oval, or too-small rectangular piece of grillwork, use it as an inset in a piece of plywood or other material.

Materials: ¾-inch plywood, iron grille, wood veneer tape, glue, nails, wood filler or putty, and the desired finish.

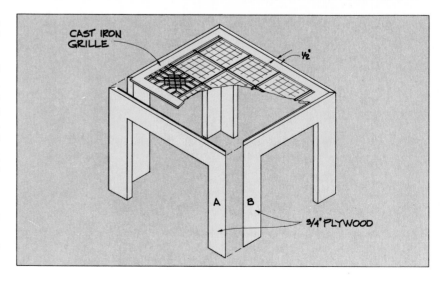

CAST IRON GRILLE

½"

A   B

¾" PLYWOOD

UP-AND-AWAY STORAGE BOX

In the battle between you and basement or garage clutter, you need all the help you can get. This serviceable unit not only provides ample room for lots of gear, but also allows you to put these items where they belong—up out of the way.

1 When planning this unit's dimensions, available space and the weight of the items to be stored are the prime considerations. You also may consider compartmentalizing the box with plywood dividers to accept specific items.

2 From ¾-inch plywood, cut the bottom of the box (A) to the desired dimensions. Cut four 8-inch-wide strips of plywood for the sides and ends (B,C). Assemble the box as shown, using glue and nails.

3 Cut a length of 1 × 2 (D) to serve as a cross brace for the box. Glue and nail it into place. If desired, install ½-inch plywood dividers. Paint the unit, if desired.

4 Cut a 2 × 4 (E) long enough to span several joists. Nail it to the bottom of the floor joists at a point where when opened the box's bottom will hit against the wall (see photograph). Nail a shorter length of 2 × 4 (F) to the joists so the front of the box will be flush with front of the 2 × 4.

5 Drill two holes in the front of the box for the rope handle. Cut a length of rope, thread the ends through the holes, and knot both ends. Screw butt hinges to the rear of the box, then with a helper, attach the box to the long length of 2 × 4. Attach the hasp hardware to the front edge of the box and to the shorter 2 × 4.

Materials: ¾-inch plywood, ½-inch plywood (optional), 1 × 2 and 2 × 4 lumber, butt hinges, hasp with rotating staple, glue, nails, length of rope, and, if desired, paint.

LOUVERED WINDOW SCREEN

This clever cover-up is ideal for a window with a less-than-desirable view and for situations in which sun control is a problem. The unit adds visual interest to the room as well.

1 The double doors shown measure 36 × 80 inches.

2 From 1 × 4 lumber, cut two uprights (A) and two crossmembers (B). Center the top crossmember over the opening and nail it securely to the ceiling joists. Nail the uprights to the crossmember and to the window casing. Add 1 × 1 cleat (C) at the bottom.

3 Cut a 1 × 2 stop (D) to size and nail it to the underside of the remaining crossmember. Recess it the thickness of the louvered doors. Position the crossmember so it in effect becomes the "top jamb" for the doors; nail into place.

4 Sand the members, then finish as desired.

5 Hinge the doors to the uprights. Add magnetic catches and door pulls.

Materials: 1 × 4, 1 × 2, and 1 × 1 lumber; louvered doors; nails; butt hinges; pulls; magnetic catches; and the desired finish.

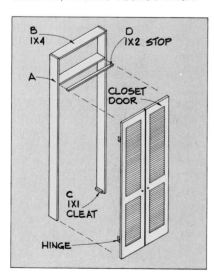

NIFTY
KNIFE BLOCK

This project offers an attractive alternative to throwing your knives back into a crowded drawer after each use. With **one of these blocks, you can store your cutlery within easy reach and protect vulnerable cutting edges, too.**

1 Tailor the size of the knife block to suit your needs. This one is 10 inches tall.

2 Cut four 1 × 6 uprights (A). Cut about a dozen strips of ⅛-inch-thick hardboard (B) to the same length as the 1 × 6s. Then clamp the 1 × 6s together and mark the intended position of the hardboard strips. Remove the clamps. Using your marks as a guide, laminate the pieces of 1 × 6 and hardboard. Use clamps to ensure a good bond. (You can vary the position of the hardboard strips to accommodate several blade widths.)

3 Cut a length of 1 × 6 (C) as long as the 1 × 6 assembly is wide, and glue it to the bottom of the uprights. Finish the unit as desired. Polyurethane varnish was applied here, but you may want to stain or paint the unit.

Materials: 1 × 6 lumber, scraps of ⅛-inch-thick hardboard, glue, and the desired finish.

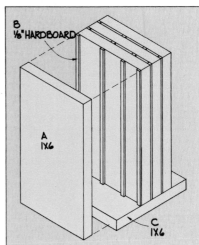

SLIP-TOGETHER PLANT STAND

If you're like most plant enthusiasts, you have a lot more plants than you have places to put them. Here's a low-cost, easy-to-construct project that provides enough display space for several of your greenery favorites.

1 The unit shown measures 18 × 36 inches and is about 16 inches tall. Alter the dimensions, if desired, to fit the floor space you have.

2 Cut six lengths of 1 × 6 (ripped to 4½ inches) for the sides (A) and four lengths for the end panels (B). Then cut the 1 × 12 shelf (C) to the desired length. Position the shelf between two of the end pieces (see sketch), and glue and nail the members together.

3 Cut ¾ × ¾-inch notches in all of the end pieces as shown. The distance between the inside edges of these notches should be 11¼ inches. Cut same-sized notches in the side members. Position them an equal distance from each end . Don't notch bottom edge of the bottom piece or the top edge of the top piece.

4 Sand and finish the plant stand as desired. Coat the unit with a water-resistant finish to minimize the possibility of water damage.

5 After the finish has dried, assemble the unit "log-cabin" style without glue or nails so the plant stand can be taken apart quickly and easily.

Materials: 1 × 12 and 1 × 6 lumber, glue, finish nails, and the desired finish.

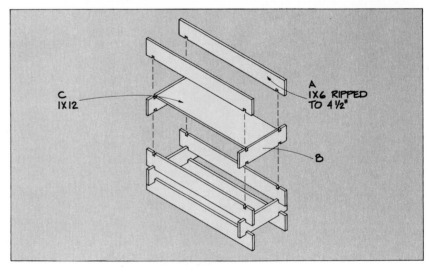

PARSONS-STYLE PLANT TABLE

If you'd like to see a little more plant and a lot less pot, you'll appreciate this Parsons-style plant table. Most of each pot is out of sight beneath the tabletop, leaving the table's surface free to accentuate the beauty of the greenery.

1 The plant table shown measures $12 \times 36 \times 18$ inches high. If desired, you can alter the dimensions of the table to fit the floor space you have.

2 On a 4×8 sheet of ¾-inch plywood, mark the outline of the various members, then cut all of the pieces (A,B,C) at one time. Be sure to allow for saw kerfs when laying out the members. Miter the edges of pieces A,B; glue and nail them together. Clamp the joints until the glue dries.

3 Cut a series of holes in the tabletop with a saber saw (the diameter of the pots you'll be using will determine the size of the holes).

4 Using glue and nails, secure the tabletop to the frame. Depending on the load, you may want to nail ledgers beneath the tabletop for additional support. With normal loads, however, this shouldn't be necessary.

5 Fill all exposed plywood edges with wood putty. Finish the unit as desired. Here, the table sports two coats of bright yellow alkyd-base paint. Sand lightly between each coat of paint.

Materials: ¾-inch plywood, glue, finish nails, wood putty, and the desired finish.

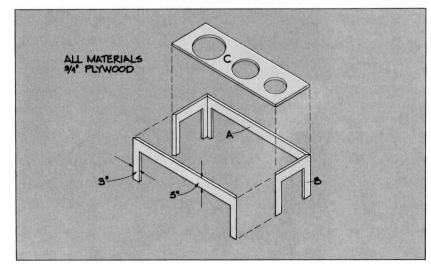

ALL MATERIALS ¾" PLYWOOD

MIRRORED SHELF UNIT

Nothing visually expands space the way a mirrored wall does. And when you add a bank of shelves to it (as shown here), the results are spectacular both visually and functionally. Though this unit is in a living room, it would work just as well in a dining room, study, or any small room.

1 This shelf unit extends from floor to ceiling, and is a bit more than 4 feet wide. If you lengthen the shelves, add more 1 × 2 facers for more shelf support. For a freestanding unit, shorten the uprights.

2 From 1 × 10 lumber, cut two uprights (A) to size. Cut grooves in the uprights to receive adjustable shelf standards. Screw the standards to the uprights.

3 Cut six 1 × 10s (B) to the length you want the shelves to be. Then cut four 1 × 3s (C) to the same length as the shelves.

4 Construct the header and bottom shelf by gluing and nailing B and C together as shown.

5 Using screws or toggle bolts, secure one of the uprights to the wall. Counterbore the screws. Then with a helper, sandwich the header and bottom shelf between the uprights. Join the members with glue and nails.

6 Cut four 1 × 2s (D) to size. Cut a groove in each to receive shelf standards along the back side. Screw standards to 1 × 2s.

7 Fill all nail and screw holes, then sand unit smooth. Paint.

8 Apply mirror tiles to the wall area using adhesive. Install shelf clips at the desired heights, and put up the shelves. Nail and glue the 1 × 2 facers (D) as shown. Support the front edge of the shelves with clips. Sink nailheads, fill, and touch up with paint.

Materials: 1 × 10, 1 × 3, and 1 × 2 lumber; shelf standards and clips; mirror tiles and adhesive; glue; nails; screws or toggle bolts; wood putty; and paint.

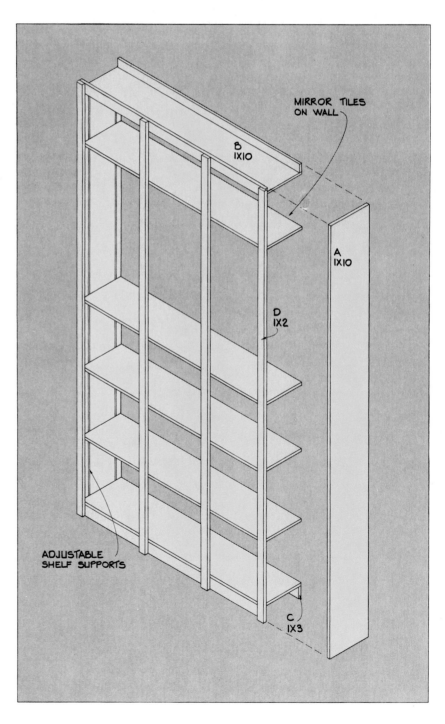

MIRROR TILES ON WALL

B 1X10

A 1X10

D 1X2

ADJUSTABLE SHELF SUPPORTS

C 1X3

DOUBLE-DUTY DIVIDER UNIT

Here's a novel way to create more coat closet space where you need it—near the entry to your home. Besides adding extra closet room, this unit helps ease another all-too-common problem—not enough shelf space.

1 The unit shown measures 30 inches wide and extends from floor to ceiling. But you can alter the dimensions to suit your particular situation.

2 From ¾-inch plywood, cut the shelving unit's uprights (A) and top and bottom plates (B). Secure the bottom plate to the floor, and the top plate to the ceiling. (Use a plumb bob to ensure the correct positioning of the plates.) Glue and nail uprights to both ends of the plates. Back the unit with a piece of ¾-inch interior plywood (C). (Please note that the plywood back fits inside the plates and the uprights.)

3 Cut four shelves (D), and position them between the uprights in the desired positions. Glue and nail them into place. (Here, two 3-inch-diameter holes were cut in the bottom shelf to accommodate umbrellas.) Cut two vertical dividers (E) to size, then glue and nail them into place.

4 Cut 1 × 4 shelf facers (F,G), then glue and nail them as shown in the sketch.

5 Measure the distance between the back of the divider and the wall, and cut a plywood shelf (H) to fit the space. Don't forget to factor in the 1 × 2 frame that surrounds the plywood. Cut the 1 × 2 frame members (I,J) to size,

then glue and nail them to the shelf. Drill a 1¼-inch hole in two 1 × 4 blocks (K) to receive a 1¼-inch wood dowel, then secure the dowel assembly to the wall and divider. (Position the dowel 63 inches from the floor.) Position the shelf a few inches above the dowel, then secure it to the wall and the back of the divider.

6 Fill all voids with wood putty or cover them with wood veneer tape. Then sand the unit smooth and finish as desired. Apply rows of mirror tiles to the ends of the unit, if desired.

Materials: ¾-inch plywood, 1 × 4 and 1 × 2 lumber, 1¼-inch wood dowel, finish nails, glue, wood veneer tape or putty, the desired finish, and mirror tiles and adhesive (optional).

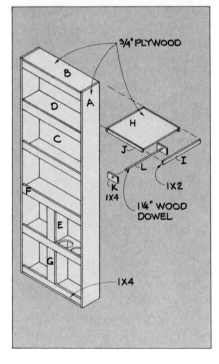

DOUBLE-DECKER PLANT STAND

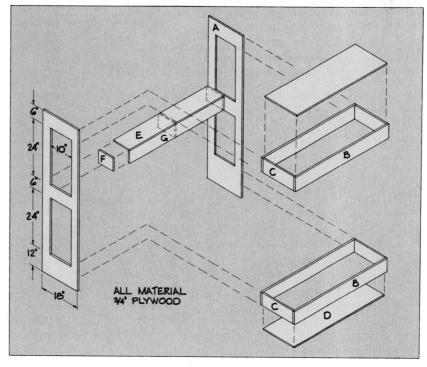

1 Plan the dimensions of your unit before beginning. This unit is 18 × 48 × 72 inches.

2 From ¾-inch plywood, cut two uprights (A). Make two rectangular cutouts, one above the other, in each upright.

3 To build the two plywood boxes, cut two sides (B), two ends (C), and a bottom (D) for each. Glue and nail the members together using butt joints.

4 Cut the shelf members (E,F,G) to size, then glue and nail these members together as shown.

5 To assemble the plant stand, turn the uprights on their sides, then sandwich the shelf and the two boxes between them. Note that the upper box is flush with the top of the uprights; the lower one should be several inches off the floor. Glue and nail the components together.

6 Fill exposed plywood edges and nail holes, sand as necessary, then paint the plant stand. A water-resistant finish will stand up best to the moisture.

7 Fasten a fluorescent light fixture with special grow-light to the inside of the top box, then run necessary wiring. Line the bottom of the lower box with a heavy polyethylene film, and fill with gravel, bark, or other suitable material. Install hooks or other plant hangers inside the top box in the positions desired.

Materials: ¾-inch plywood, nails, glue, paint, fluorescent light fixture (with a tube designed to promote plant growth) and wiring, polyethylene film, and plant hangers and screw hooks.

ALL MATERIAL ¾" PLYWOOD

MODULAR TELEPHONE CENTER

With this stylish duo, you'll find that talking by telephone can be as comfortable as it is enjoyable. Both units measure 18 inches square and are made from two layers of sturdy ¾-inch plywood.

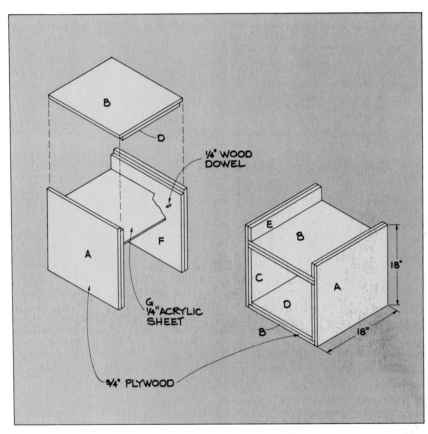

1 To construct the seat unit, first cut two 18-inch-square pieces (A) for the sides. Next cut an 18 × 16½-inch piece (B) for the base. Assemble A and B using glue and nails. Cut the remaining seat unit members (C,D,E) to size (piece E measures 18 × 3 inches), then assemble as shown using glue and nails.

2 For the phone stand, cut two more 18-inch square pieces for the sides and another 18 × 16½-inch piece for the base. Glue and nail together. Then cut the remaining members (D,F) to size; assemble remainder of frame.

3 Drill two ¼ × ¾-inch-deep holes in each side of the stand's legs for the shelf (G), which is a piece of ¼-inch acrylic measuring 18 × 15 inches. Insert a 1½-inch length of ¼-inch dowel in each hole, and glue in place.

4 Fill all voids. Sand smooth, then finish both units.

5 Slide the acrylic shelf in place, and top the seat with an upholstered 3-inch-thick foam cushion (see page 96).

Materials: ¾-inch plywood, ¼-inch acrylic sheet, ¼-inch wood dowels, glue, nails, wood putty, foam cushion, upholstery fabric, and desired finish.

BOLT-TOGETHER ARMCHAIR

Using this simple design, you can turn out a whole roomful of chairs in no time—and at a price that's much lower than you might imagine. Lengths of pine lumber form the sides and legs, and a piece of plywood serves as the seat.

1 The plywood seat of the chair shown measures 24 inches square, and the legs are 27 inches tall. You can widen the unit, if desired, to create a sofa or love seat, but the distance from the floor to the seat should remain about 16 inches, including the cushion thickness.

2 Cut four 2 × 2 legs (A) to size. Cut four crossmembers (B) from 1 × 8 lumber; these should be as long as the plywood seat is deep. Drill and bolt the crossmembers to the legs as shown. Position the top edge of the lower crossmember at finished seat height, about 16 inches from the floor.

3 Cut 1 × 8 stretchers (C) to size (these should be 4½ inches wider than the seat will be), then bolt them to the chair legs. Position the back two stretchers at the same height as the crossmembers. Position the front stretcher so its top edge is 1 inch above the seat bottom (F).

4 Cut, then nail 1 × 2 ledgers (D,E) to the crossmembers and stretchers as shown.

5 Cut the ½-inch plywood seat platform (F) to size. Glue and nail it to the ledgers.

6 Sand the entire unit smooth and finish it as desired. Cut a square of 5-inch foam to fit the plywood seat, and upholster it in the fabric desired (see page 96). Add a group of throw pillows for the backrest.

Materials: 2 × 2, 1 × 8, and 1 × 2 lumber; ½-inch plywood; glue; nails; bolts; washers; nuts; 5-inch-thick foam slab; upholstery fabric; and finish.

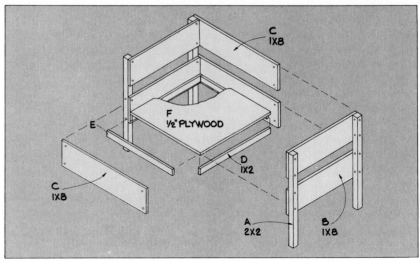

SOFA AND END-TABLE COMBO

This strictly contemporary unit is both functional and flexible. It has seating space galore, **plus ample end table surface. It even doubles as a bed when overnight guests arrive.**

1 Each of the modules measures 30 × 72 inches. You can tailor the dimensions if desired.

2 Cut all members (A,B,C) to the desired size. Assemble as shown, using glue and nails. Strengthen the corner joints, if desired, with metal corner braces. Fill voids in plywood, sand, and finish units.

3 Top units with upholstered cushions and pillows (see page 96).

Materials: 1 × 12 lumber, ¾-inch plywood, foam cushions and pillows, fabric, glue, nails, wood putty, and finish.

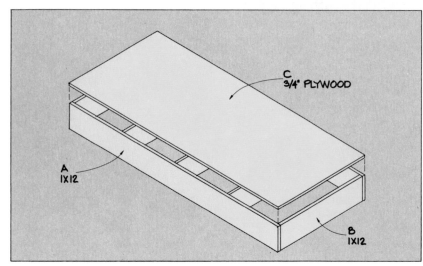

C
3/4" PLYWOOD

A
1X12

B
1X12

ALL-IN-ONE SOUND CENTER

Even though it occupies only about 3½ feet of wall space, this good-looking component center is roomy enough to put **everything in its place. The lower shelves have drop-down doors that conceal their contents and keep out dust.**

1 The unit shown is 7 feet tall. Adjust the size and position of the shelves to suit your needs.

2 Cut 1 × 4 and 1 × 3 uprights (A,B) to size. Then cut cross braces (C) from 1 × 4 lumber. Glue and nail the side assemblies together as shown. Maintain equal spacing between the vertical members.

3 For the unit's back (D), cut a piece of ¾-inch plywood to the desired size. Glue and nail the side assemblies to plywood back.

4 Cut the ¾-inch plywood shelves (E,F) to size. Note that the upper three shelves are recessed only slightly, whereas the lower three sit back about 2 inches from the front of the unit.

5 Position the shelves at the desired heights and sandwich them between unit's side assemblies. Secure with glue and nails.

6 Cut a 1 × 8 top brace (G) to size, then glue and nail into position as shown.

7 Cut doors (H,I) for the lower shelves from ¾-inch plywood. Size these carefully so they will be able to clear the shelf at their upper edge. Also take into account the width of the washers used as part of the hinge mechanism (see detail).

8 Cover all exposed plywood edges with veneer tape, sand unit smooth, and apply finish.

9 When the finish is dry, hold the doors in their closed position and drill pilot holes for the screws. Enlarge the holes in H to accommodate a piece of plastic tubing. Hinge as shown in detail.

10 Add door pulls, and eye screws and sash chains for sup-

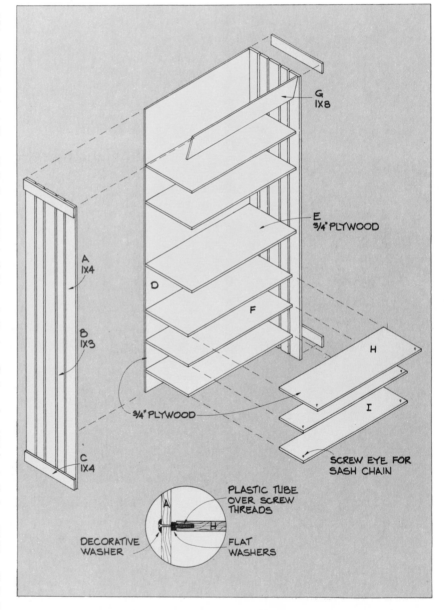

port. Secure the unit to the wall studs using wood screws.

Materials: ¾-inch plywood; 1 × 8, 1 × 4, and 1 × 3 lumber; wood screws; decorative and flat washers; screw eyes; sash chain; glue; finish nails; wood veneer tape; plastic tubing; door pulls; and desired finish.

WALL-SPANNING WINE RACK

This highly unusual project not only allows you to show off your store of wines to good advantage, but also functions as a buffet surface that's ideal for serving. The glass top and shelves add a light and lively flavor to the unit.

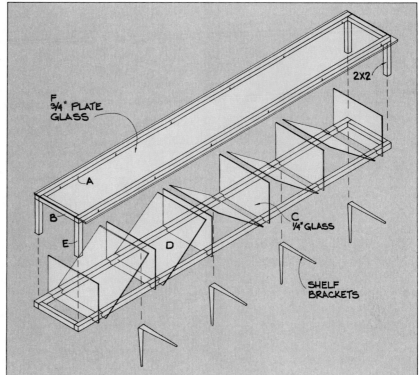

1 The unit shown extends along the length of one wall and is about 14 inches from top to bottom. But feel free to alter the dimensions to suit your situation.

2 Cut the frame members (A,B) from lengths of 2 × 2 lumber. Glue and nail together as shown. Sink the nails. Position both frames side by side and mark the positions for the vertical glass dividers (C). Cut ¼-inch-wide, ¾-inch-deep dadoes to receive the verticals. Also cut notches in the lower frame as shown in the sketch to accept the diagonal dividers (D). These simply lie against the vertical dividers.

3 Join the upper and lower frames by sandwiching 2 × 2s (E) between them. Secure with glue and nails. Sink nails.

4 Sand the unit smooth and finish as desired. Then with a helper fasten the wine rack to the wall by driving screws through the 2 × 2 framework into the wall studs. Further strengthen the rack by screwing shelf brackets to the framework and wall studs.

5 Insert the shelves into the dadoes and notches as shown. Top the unit with a piece of ¾-inch-thick plate glass (F).

Materials: 2 × 2 lumber, shelf brackets, ¼-inch tempered and ¾-inch plate glass with seamed edges, finish nails, screws, glue, and desired finish.

FULL-DAY AND WEEKEND PROJECTS

If you're like most people, you don't mind spending some time working on a project as long as the results are worth it. In this chapter is a grouping of special projects worth every bit of effort you put into building them.

A glance through the following pages will reveal a classy collection of tables (all types—dining, end, coffee, and others), buffets, storage units, bookshelves, work centers, and many other equally serviceable units. And you can build any of these handsome pieces in a weekend or less.

If questions pop up while you're working on a project, don't forget about the "Planning and Building Basics" section, which begins on page 84. Maybe you're unsure about how to make a miter cut, fasten a base to a cabinet, anchor a shelf unit to a wall, or apply a particular finish to a project. Whatever the problem or situation, chances are good that the basics section has the answer for you, explained in easy-to-understand terms.

CLASSIC-LOOK TRESTLE TABLE

Trestle tables have long been valued for their straightforward good looks and sturdy construction. Our version continues these traditions in fine style. Wrought iron braces add both stability and decoration to the table.

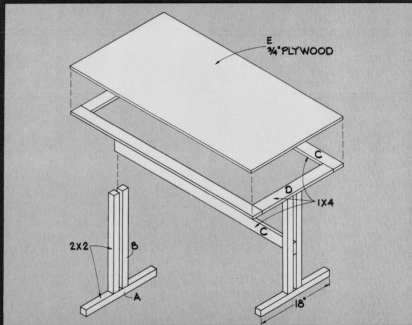

E ¾" PLYWOOD
C
D
1X4
C
2X2
B
A
18"

1 The table shown measures 24 × 48 inches and is 30 inches tall. If desired, you can alter the dimensions to suit your needs.

2 From 2 × 2 lumber, cut two base pieces (A) and four uprights (B). From 1 × 4 lumber, cut the remaining frame members (C,D). (Size these 1 × 4s so the tabletop will overlap the frame ½ inch all around.) And finally cut the tabletop (E) from ¾-inch plywood.

3 Sandwich the stretcher (C) between the 2 × 2 uprights. (Position the stretcher carefully so it will be level and perpendicular to the uprights.) Secure as shown using glue and nails. Sink the nails. Turn the leg assembly upside down, and nail and glue the base members (A) as shown. Naturally, you'll want to center these members.

4 Assemble the tabletop frame members (C,D) as shown using glue and counterbored screws. Fill recesses with dowel plugs. Turn the leg assembly right side up, then glue and screw the frame to the uprights. Invert the assembly again, and center it over the plywood tabletop. Secure with glue and screws.

5 Finish exposed plywood edges with wood veneer tape. Do any necessary sanding, fill all nail holes, and apply finish.

6 For extra support, fasten wrought iron braces as shown.

Materials: ¾-inch plywood, 2 × 2 and 1 × 4 lumber, eight wrought iron braces, glue, finish nails, wood screws, dowel, wood veneer tape and contact cement, wood putty, and desired finish.

CLEAN AND CONTEMPORARY BUFFET

Leather drawer and door pulls accent this most impressive natural-finish unit. The cabinet shell is good-quality plywood and 1×2s; mitered edges add to the clean look. Add shelves inside to better organize the storage space.

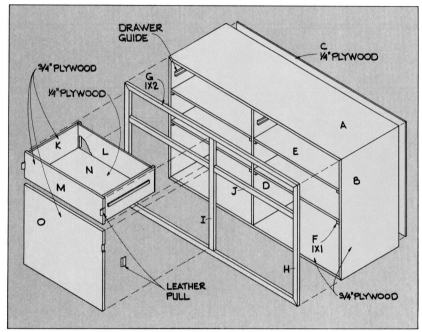

1 The unit shown measures 18 × 48 × 30 inches.

2 Start by cutting top and bottom pieces (A) and two sides (B) from ¾-inch plywood. Miter-cut the ends of each piece. Then cut ⅜ × ¼-inch rabbets in the back edge of each member. Nail and glue the members together.

3 Cut the back (C) from ¼-inch plywood, then nail it to the cabinet frame members (A,B).

4 From ¾-inch plywood, cut a vertical divider (D) and four shelves (E). Then cut eight 1 × 1 ledgers (F). Position and secure the divider, then glue and nail the ledgers to the frame and divider at the desired height. Position the shelves, and secure with glue and nails.

5 Trim the cabinet front with 1 × 2s (G,H,I,J) as shown.

6 Cut drawer members (K,L,M,N) using ¾-inch and ¼-inch plywood. Cut ¼ × ⅜-inch grooves in members K,L, and M to accept the drawer bottom. Cut notches in the drawer fronts for the leather handles. Assemble drawers using glue and nails. Fasten drawer guides as shown.

7 Cut the cabinet doors (O) to fit. Notch the doors (see sketch). Hinge the doors to the cabinet.

8 Sink all nailheads, fill all recesses with wood putty, sand the unit smooth, and finish as desired. When the finish dries, secure door and drawer pulls.

Materials: ¾- and ¼-inch plywood, 1 × 2 and 1 × 1 lumber, drawer guides, leather strips, hinges, wood glue, wood putty, finish nails, and desired finish.

FREESTANDING HOBBY CENTER

1 Plan the dimensions of the project to suit your needs. This unit is 60 × 16 × 36 inches. To use it as a desk, plan the work surface to be 30 inches tall.

2 For the work surface, cut two plywood pieces (A,B) so the top piece is ¾ inch longer than the bottom one. Secure together as shown with glue and nails. For the "leg," cut two pieces (C,D) so piece D is ¾ inch longer than piece C. Again laminate the pieces as shown with glue and nails.

3 Secure the work surface to the leg using glue and nails. Fortify the joint by screwing a metal angle to both members. Cover the assembly with plastic laminate (see page 93).

4 For the shelf unit, cut two sides (E), a top and bottom (F), and a shelf (G) from ¾-inch plywood. Cut ¾ × ⅜-inch dadoes in the sides, then glue and nail the members together as shown. Cut four base pieces (H,I) from 1 × 4 lumber, then assemble using glue and nails. Secure the base assembly to the shelf unit. Fill voids in plywood, sand the unit smooth, and paint.

5 To secure the work surface to the shelf unit, first cut a length of 1-inch-diameter steel pipe to span the distance between the shelf and the work surface. Secure into place with two pipe flanges.

Materials: ¾-inch plywood, metal angle, plastic laminate and adhesive, 1 × 4 lumber, 1-inch steel nipple, pipe flanges, wood putty, glue, nails, and paint.

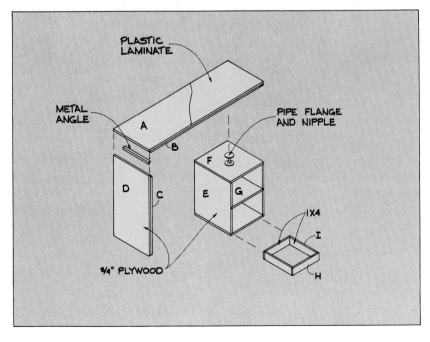

SMART AND SASSY COFFEE TABLE

Ever noticed how coffee tables just seem to attract clutter? It's almost as if they were magnetized. That's not a problem with this version, though. As books and magazines accumulate, just stash the older ones in the handy storage cubicles.

1 Plan the dimensions of the coffee table (this one is 24 × 60 inches) so it is in proportion with the rest of your furniture. Standard coffee table height is 15 to 17 inches.

2 Start by cutting a piece of ¾-inch plywood (A) the size you want the tabletop. Miter-cut four pieces of good-quality 1 × 2 (B,C) to frame the plywood. Bevel and smooth the edges of the 1 × 2s for a finished look. Glue and nail the 1 × 2s to the tabletop so the plywood extends ⅛ inch or so above them.

3 For the storage units, cut pieces of ¾-inch plywood (D,E). Glue and nail together. Cut doors (F) for each end from ½-inch plywood, and frame them with more 1 × 2s (G,H), miter-cut, beveled, and sanded smooth. For the storage cubicles' bases, cut lengths of 1 × 2s (I,J), then glue and nail together. Secure the bases to the bottom of the cubicles.

4 So the storage units' doors will swing freely, cut scrap plywood blocks (K) and fasten them to the cubicles.

5 Sand any rough surfaces, then finish as desired. Fasten the doors to the storage units using continuous hinges. Add magnetic catch hardware. Assemble the coffee table by driving wood screws up through the storage units into the tabletop.

Materials: ¾- and ½-inch plywood, 1 × 2 lumber, continuous hinges, magnetic catch hardware, glue, finish nails, wood screws, and finish.

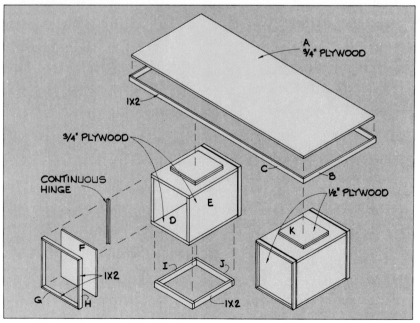

RAINBOW-INSPIRED COFFEE TABLE

Let your imagination run free when it comes time to finish this exciting table creation. Here, ceramic tiles of various hues create a splash of color that makes the table an instant attention-getter. The base is made from lengths of hollow-core door.

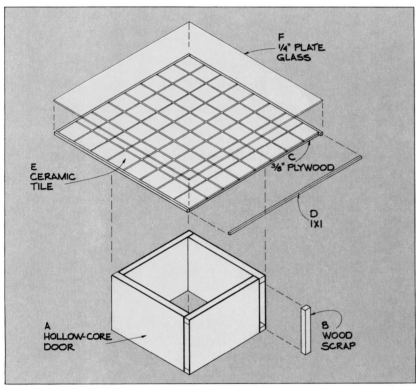

1 Plan the tabletop dimensions carefully. To ensure an accurately sized frame, position the ceramic tiles as they will lie, then measure the length and width.

2 Cut a 12 × 80-inch hollow-core door into four equal pieces (A). Plug the cut ends of the door with wood scraps (B); glue into place. Glue and nail the base pieces together as shown. Finish all exposed edges with wood veneer tape.

3 Cut a piece of ⅜-inch plywood the same size as the ceramic tile measurements to serve as the base for the tabletop (C). Cut lengths of 1 × 1 lumber (D) to fit around the edges of the plywood; glue and nail them to the plywood. Position 1 × 1s so tiles will be flush with top of frame. Glue and nail tabletop to base.

4 Apply adhesive to plywood, and lay the tiles (E). Allow 24 hours to dry, then apply grout. Seal the grout with sealer.

5 Sink nails, fill all nail holes, then sand all edges and surfaces smooth. Finish as desired. Top with ¼-inch plate glass (F).

Materials: ⅜-inch plywood; 12 × 80-inch hollow-core door; 1 × 1 lumber; wood scraps and veneer tape; ceramic tiles, adhesive, grout, and silicone sealer; ¼-inch plate glass; finish nails; glue; wood putty; and desired finish.

ROLL-AROUND SERVING CART

With this sleek, two-tiered server, dispensing snacks is easy. And when you aren't using the unit, put it to work as a display case for knicknacks and other decorative items.

1 Before beginning construction, plan the dimensions carefully. The unit shown here measures 17½ × 32¾ × 28¼ inches.

2 Start by building triangular frames (A,B,C). Piece A is 43¼ inches long. Angle-cut 2 × 2s; assemble with glue and nails.

3 Clamp the frames together, then drill holes where shown to accommodate 1-inch dowels (D).

4 Cut five pieces of dowel, two at 21 inches long, and three at 19 inches, then thread them through the previously drilled holes to join the frames. The distance between the inside edges of the frames should be 14½ inches.

5 Cut the plywood shelves (E,F) to size, then cover their top surface with plastic laminate.

6 Trim the shelves with 1 × 2s (G,H) secured flush with the bottom of the plywood shelves. Position the shelves atop the dowels; nail the shelves to frames.

7 Cut four 6-inch wheels (I) from ¾-inch plywood, slip them onto the dowels, and secure with ¼-inch dowels as shown.

8 Fill all exposed plywood edges and voids with wood putty, then sand and finish the unit as desired.

Materials: 1 × 2 and 2 × 2 lumber, ¾-inch plywood, 1- and ¼-inch dowel, glue, finish nails, plastic laminate and adhesive, wood putty, and finish.

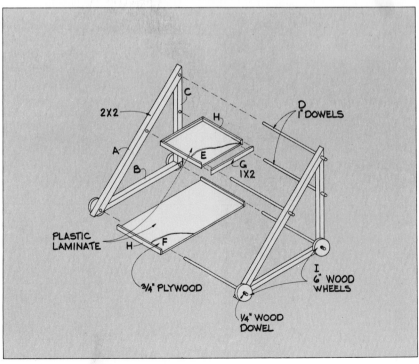

SLIDE-APART COFFEE TABLE AND BAR

Were it nothing more than a good-looking piece of furniture, this unit still would be well worth its cost. But it is more. The table's base serves as a convenient stash for beverage containers and other bar supplies.

1 The project shown measures 30 × 42 × 15 inches, but you can adjust these measurements.

2 From ¾-inch plywood, cut pieces for the table base (A,B,C). Miter members A and B. Assemble using glue and nails.

3 The tabletop assembly consists of two open-sided boxes with wood guides attached to the sides. To construct each box, first cut the members (D,E,F) from ¾-inch plywood, then miter the edges as shown. Glue and nail all pieces together.

4 Cut four lengths of 1 × 3 (G) for wood guides. Center and cut a ¾ × ⅜-inch-deep groove along the length of each piece to receive the 1 × 1 wood strip (H). Using the dimensions of the table base as a guide, secure the wood guides to the underside of the tabletop assemblies with glue and nails. Allow ¹⁄₁₆-inch tolerance between gliding parts.

5 Cut lengths of 1 × 1 for the guide strips (H). To position the strips, measure the distance from the underside of the top assemblies to the top of the groove in the wood guides. Allow for clearance, and draw a line along the side of the table base at this point. Glue and nail wood strips to the base. Add a 1 × 6 shelf (I) to inside of base as shown.

6 Sink nails, fill voids with wood putty, and cover exposed plywood edges with veneer tape. Finish as desired.

Materials: ¾-inch plywood; 1 × 6, 1 × 3, and 1 × 1 lumber; glue; finish nails; wood veneer tape; and the desired finish.

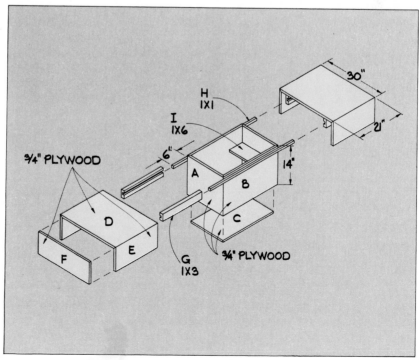

ADJUSTABLE-HEIGHT FLOOR LAMP

If you always seem to have trouble getting a light source exactly where you want it, this striking lamp will make your life a lot easier. And its unusual design is sure to make it one of your most-talked-about furnishings in any room.

1 Plan the lamp's dimensions carefully. The unit's upright assembly measures 60 inches.

2 To construct the arm assembly, first cut two short 1 × 3 blocks (A). Cut one end of each as shown. Then cut two 1 × 2s (B) to size, cutting both ends of each piece to a point. Laminate all arm pieces together with glue.

3 Drill three ⅝-inch holes at the lamp end of the arm. Drill an additional hole for the stationary dowel and one about ⅓ of the way up the arm.

4 For the upright portion, cut two lengths of 1 × 3 (C). Cut one end of each to a point. Cut two 1 × 4 blocks with similar points at both ends (D), one 1 × 3 block (E), pointed on one end, and three 1 × 3 spacer blocks (F)—cut one of them short so wiring can pass through. Drill holes along the length of C. Laminate C,D,E, and F as shown.

5 For the base, cut an 18-inch triangular frame of 1 × 3s (G), two ½-inch plywood triangles (H), and one ⅜-inch-thick slate triangle (I). Or if desired, cut two plywood triangles. To add weight, use pea gravel to fill the space between the plywood when assembling the base. Notch members as needed to allow for running the wiring later. Cut a rectangle from the plywood triangles and slate to receive the upright portion of the lamp. After switch and electrical cords have been routed, assemble the base and glue and nail the upright to the base. Trim opening with quarter round (J,K).

6 Cut four lengths of ⅝-inch

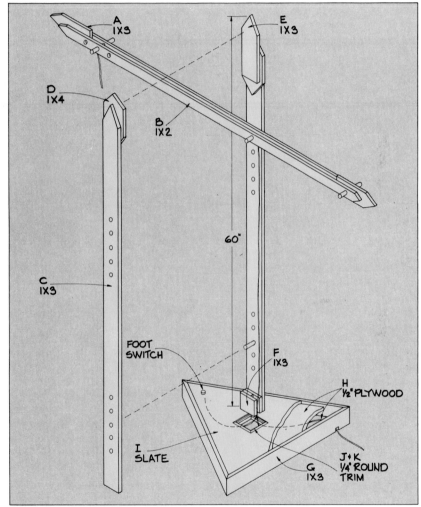

dowel. Cut a length of chain to hold the arm. Attach a metal ring to each end of the chain.

7 Finish all wood members as desired. Let dry completely.

8 Position the arm portion of the lamp, then insert all of the dowels as shown. Rig the chain, looping the rings over the stationary dowel pin and the adjustable pin in the upright.

9 Either purchase a lighting fix-

ture or fashion one yourself. Hang the lamp from the dowel pin at the end of the arm support.

Materials: 1 × 4, 1 × 3, and 1 × 2 lumber; ⅝-inch dowel; ½-inch plywood; quarter round; chain; metal rings; ⅜-inch-thick slate (optional); lighting fixture; electrical cord with plug; foot switch; wood glue; finish nails; and the desired finish.

OVERHEAD POT AND PAN RACK

All too often, kitchen ceiling space goes to waste. Why not put yours to good use with this sturdy rack and shelf combi- **nation? It will hold almost all of your pots and pans, and its rugged good looks will add architectural interest, too.**

1 The rack shown hangs down about 26 inches from the ceiling and measures 36 × 60 inches.

2 Cut four 4 × 4 uprights (A). Flute the edges and chamfer the lower ends. Cut rails (B,C) from 2 × 4s. Bolt longer rails (B) to the top ends of the uprights, and glue and nail the shorter ones (C) between longer rails.

3 Cut 2 × 4 bottom rails (D) and stretchers (E). Nail stretchers between the lower rails, 12 inches in from each end; then bolt D,E to the uprights.

4 Cut 2 × 6 shelf members (F) to size, then nail them to frame members. Notch 2 × 6s as needed.

5 Cut several 3-inch lengths of ¾-inch dowel. Drill holes in the lower rails and stretchers for them. Glue dowels into holes.

6 Sand the unit smooth, seal, and finish as desired. After the finish dries, if you plan to incorporate lighting into the unit, secure the fixtures between lower rails and run wiring.

7 To secure the unit, first determine the direction and position of the ceiling joists. Then with a helper, lift the rack to the ceiling (at a right angle to the ceiling joists) and mark the location of the holes you will later drill in the top rails (C).

8 Drill holes in the top rails. Lift unit into position and screw it to joists. Complete wiring.

Materials: 4 × 4, 2 × 6, and 2 × 4 lumber; ¾-inch-diameter wood dowel; screws; glue; bolts; washers; nuts; finish nails; light sockets and wiring (optional); and desired finish.

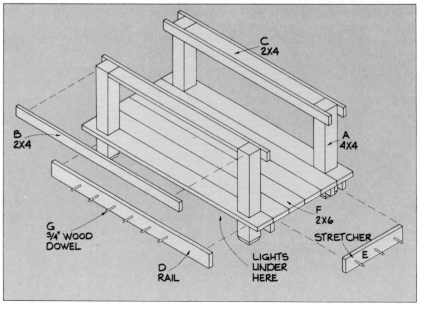

ROLL-AROUND MICROWAVE CART

Most kitchens simply weren't designed to accommodate a microwave oven. And though putting a microwave on a counter will work, it means a loss of valuable work surface area. Here's a sensible alternative that not only houses the oven but also can store utensils and cookbooks beneath.

1 When planning the width and depth of the cart, let the dimensions of the microwave dictate its size. Allow at least 1 inch headroom above the oven for ventilation. The height should be 30 to 36 inches, 36 being the standard countertop height.

2 From ¾-inch plywood, cut the unit's two sides (A). When cutting these pieces, be sure to keep in mind that the casters will add 3 inches to the unit's height.

3 Cut the bottom (B) and three shelves (C) from ¾-inch plywood. Cut three 1 × 2 stretchers (D) the same length as C. Using butt joints, glue and nail the bottom and two lower shelves into place. Use 1 × 2 ledgers (E) to strengthen the shelf the microwave will sit on. Glue and nail the three 1 × 2 stretchers to the underside of the top shelf; glue and nail the assembly to the cart.

4 Fill voids in plywood with putty; smooth and finish the unit as desired. Cover the work surface with plastic laminate (see page 93).

Materials: ¾-inch plywood, 1 × 2 lumber, plastic laminate and contact cement, 3-inch plate-type ball casters, glue, finish nails, screws, wood putty, and finish.

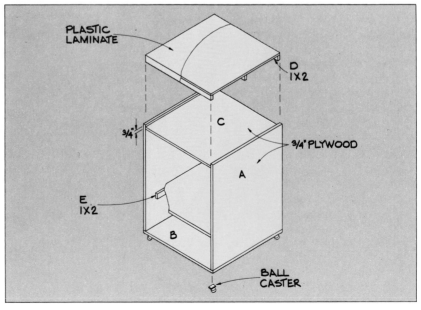

PLASTIC LAMINATE

D
1X2

C

¾"
¾"

¾" PLYWOOD

A

E
1X2

B

BALL CASTER

WALL-HUNG CURIO SHOWCASE

This inventive—and relatively demanding—project takes its design cue from the classic dollhouse. Each "room" provides an ideal stage for highlighting a favorite collectible—in this case a piece of miniature furniture.

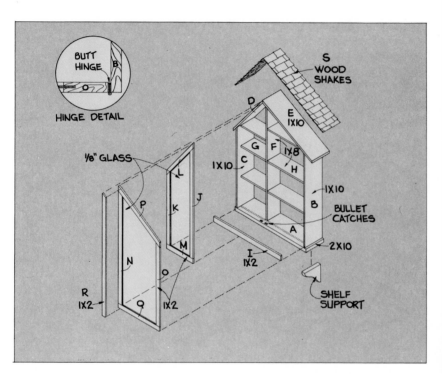

1 The cabinet shown is 38 inches wide and 56 inches tall. Adjust the dimensions to fit available wall space and your needs.

2 Fashion the 2 × 10 base (A).

3 Cut the cabinet's uprights (B,C) from 1 × 10 lumber. Cut 1 × 10 rafters (D,E). Angle-cut the members as required. Then construct the unit's outer shell. Cut 1 × 8 vertical divider (F) and shelves (G,H); nail them between the uprights.

4 Cut facer (I) from 1 × 2 lumber. Nail and glue it between the cabinet uprights as shown. Cut the door frames from pieces of 1 × 2 lumber (J through Q). Before assembling, cut a ¼ × ⅜-inch-deep groove in the edges of each frame piece to receive the glass front. Miter the members as needed and assemble the outer side, bottom, and sloping top piece. Slide ⅛-inch glass into the grooves and attach the remaining frame piece. Nail a 1 × 2 stop (R) to one door.

5 Nail wood shakes (S) to the roof. Sand all edges and surfaces smooth, and finish as desired. Attach hinges and latch hardware.

6 Secure premade wood shelf supports to the wall studs, then set the project up on the supports. Tie the unit to the wall studs using corner braces.

Materials: 2 × 10, 1 × 10, 1 × 8, and 1 × 2 lumber; ⅛-inch-thick plate glass; wood shakes; hinges; bullet catches; glue; nails; shelf supports; and finish.

CUSTOM-MADE SIDE TABLE

If you like the custom-made look in furniture, you are going to appreciate this unit. Nicely tailored, yet inexpensive to build, it will serve as a table or, **if the need arises, as a study desk. This design adapts well to other furniture pieces, so you may want to build an ensemble.**

1 The table shown measures 18½ × 66 × 30. If necessary, adjust the dimensions to fit the space you have available and your needs.

2 Cut the ¾-inch plywood or particleboard tabletop (A) to size. Be sure to make your cuts carefully so the edges are all square.

3 For each base assembly, cut four plywood rectangles (B). Miter the edges at 45 degrees. Glue and nail the members together. If desired, brace the inside corners with scraps of 2 × 2 or with metal corner braces.

4 Position the two base assemblies under the tabletop. With glue and finish nails, secure the tabletop to the bases.

5 Spread contact cement on the tabletop and on the underside of a sheet of plastic laminate. Allow both surfaces to dry, then lay down the plastic laminate. For detailed instructions on how to apply plastic laminate, refer to page 93. Let dry, then trim the excess.

6 Cut 1 × 6 tabletop frame members (C,D) to size, mitering the ends. Glue and nail the 1 × 6s to the tabletop. (*Note:* If the unit will be used as a study desk, use 1 × 3s or 1 × 4s for the frame to allow for more leg room.)

7 Set and fill all nail holes, and finish the table as desired.

Materials: 1 × 6 lumber, ¾-inch plywood to match, scraps of 2 × 2 or metal corner braces (optional), plastic laminate and contact cement, wood glue, finish nails, and desired finish.

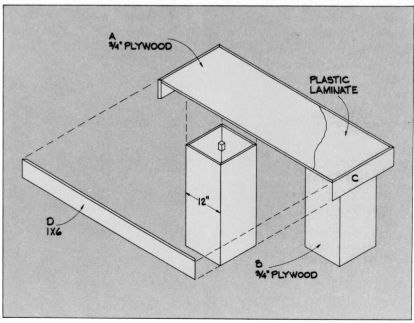

OAK PLYWOOD ARMCHAIR

This plywood chair, a great first furniture project, goes together more quickly than you might imagine. Cut side and back pieces from classy-looking oak plywood, and cover the seat and back cushions with vinyl fabric.

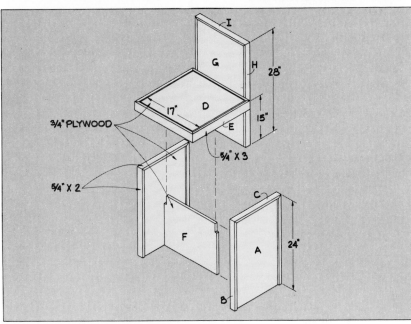

1 Begin by cutting two arm uprights (A) from ¾-inch plywood. Then cut and miter the ends of ⁵/₄ × 2 oak (B,C) to frame the uprights. Glue and nail the oak to the top and sides of the arm uprights as shown.

2 Next cut a square seat (D) from ¾-inch plywood. Cut and miter the ends of ⁵/₄ × 3 oak (E) to frame the seat. Glue and nail the oak to the seat, as shown.

3 From ¾-inch plywood, cut a seat support (F) that spans the width of the seat platform and puts the seat at the proper height. Cut a notch in each side to receive the seat frame.

4 For the seat back, cut a piece of ¾-inch plywood (G) to size. Trim with ⁵/₄ × 2 oak (H,I).

5 To assemble the chair, position the seat support as shown, then glue and screw the seat support to the arm uprights. Center the seat assembly atop the support and bolt the arm uprights to the seat trim, making sure the seat is level with the floor. Bolt the seat back to the rear seat trim. Sand all edges and surfaces smooth, stain, and finish.

6 Using 2-inch-thick foam and vinyl fabric, cut and upholster a seat and back cushion (see page 96). Secure them to the chair using nylon hook-and-loop fastening tape.

Materials: ¾-inch oak veneer plywood, wood veneer tape, ⁵/₄ × 2 and ⁵/₄ × 3 oak, wood glue, finish nails, wood screws, lag bolts, 2-inch-thick foam, upholstery fabric, nylon hook-and-loop fastening tape, and finish.

MODULAR SEATING SYSTEM

With this shapely system, you get not only attractive seating, but also a generous helping of end table space. Plate glass mirror and durable plastic laminate combine to give the unit a look perfect for modern decors.

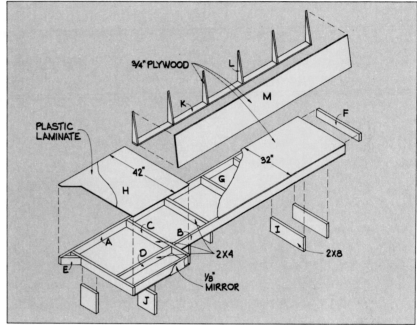

1 Before beginning construction, plan the dimensions of the unit. Comfortable seating height is between 14 and 16 inches (including cushion thickness).

2 Cut the 2 × 4 platform framing members (A through F) to length. Then glue and nail together.

3 Using ¾-inch plywood, sheathe the framework with panels (G,H) cut to size and positioned as shown. Use glue and nails to secure the panels. Cover the bottom of the frame first, then cut 2 × 8 supports (I,J) and glue and screw them into position. Glue and nail the top G,H members into place.

4 To construct the sofa back, cut a 4-inch-wide strip of ¾-inch plywood (K) to fit along the back of the seating portion of the unit. Then cut several braces (L) from plywood. Make them 4 inches wide at the bottom and 15 inches high. Glue and nail the braces to the strip. Secure the sofa back to the sofa with glue and screws. Face the sofa back members with ¾-inch plywood (M).

5 Cover the end table portion of the unit with plastic laminate (see page 93). Then finish the support members (I,J), as well as the rest of the platform. Face the front of the unit with polished plate glass mirror. Upholster cushions for seat and back (see page 96).

Materials: ¾-inch plywood, 2 × 8 and 2 × 4 lumber, glue, screws, nails, plastic laminate and contact cement, ⅛-inch polished plate glass mirror and adhesive, finish, shredded foam, and upholstery fabric.

PASS-THROUGH SNACK BAR

Handy pass-throughs like this one are common features in older homes. You can use the same idea between an addition and an existing room, or start from scratch and create an opening between rooms. This handsome treatment adds counter and eating space, and the wall-hugging cabinetry gives extra utility.

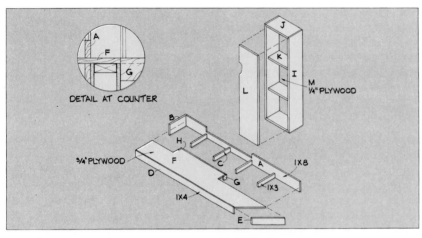

1 Begin by planning the dimensions of your project. The counter shown here measures about 7½ feet long and is 42 inches high.

2 Cut the desired opening in the wall. Install a header to form the top jamb, and a sill plate; the wall studs will serve as side jambs. Trim the opening with jamb stock and casing molding. From 1 × 8 lumber cut ledgers (A,B), butting the corners where they join. Notch A as needed to fit around the opening. Cut 1 × 3 counter supports (C) to size as well as two 1 × 4 counter facers (D,E) and a piece of ¾-inch plywood for the countertop (F). Round the edges of the facers for a sculptured look. On a level surface, glue and screw the counter assembly together. Countersink screwheads and insert dowel pins. Lift the counter into place, then secure unit to the wall studs. Trim countertop, which extends into adjacent room, with 1 × 4s (G,H).

3 Decide on the number and size of cabinets you need, then cut the sides (I) and the top and bottom (J) from good-quality boards (1 × 6s and 1 × 8s were used here). Round the edges. Cut ¾ × ⅜-inch rabbets in each side member for the top and bottom. Also cut a ¼ × ⅜-inch rabbet in the inside back edges of the sides, top, and bottom to accommodate the cabinet back.

4 Cut a cabinet back for each cabinet from ¼-inch plywood. Assemble the cabinets using glue and nails. Cut shelves (K) from 1-inch boards, then position; nail them into place. Sink nails.

5 Using 1-inch boards, cut cabinet doors (L). Cut a 1½-inch radius in each for a door pull. Hinge doors to cabinet frames.

6 Fill all voids with wood putty, sand all surfaces smooth (round corners and edges), and finish.

7 Position cabinets and secure them to the wall studs or wall.

Materials: ¾- and ¼-inch plywood; 1 × 8, 1 × 6, 1 × 4, and 1 × 3 lumber; glue; screws; nails; wood putty; dowel; jamb stock; casing molding; finish; and hinges.

GOOD 'N' BASIC WORKBENCH/ TOOL RACK

It's easy to make order out of workshop chaos with this handy combination. Large tools and bulky supplies fit **neatly beneath the work surface, and hand tools hang within easy reach from pegboard hooks.**

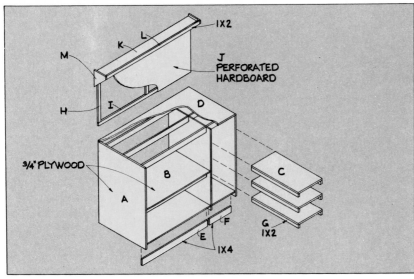

1 Plan the dimensions, keeping in mind that standard workbench height is 36 inches and that, for best use of plywood, the depth of the workbench should be just shy of 24 inches.

2 For the workbench, cut three uprights (A) from ¾-inch plywood. Plan shelf arrangements, and cut the number of larger shelves (B) and smaller shelves (C) you want. Cut a plywood top (D). Cut three strips of 1 × 4 (E) to fit beneath wider shelves B and under the top. Cut three shorter 1 × 4 strips (F) to fit under the bottom shelf of the narrow storage area, and three matching strips to support the top. Cut two 1 × 2 cleats (G) for each narrow shelf.

3 Assemble the workbench as shown using glue and nails or screws.

4 For the tool rack, cut strips of 1 × 2 (H,I) for the frame; glue and nail together. Cut a piece of perforated hardboard (J) to size, then nail it to the frame. Cut a 6-inch-wide shelf (K) from plywood, and a front 1 × 2 rail (L). Glue and nail these members together, making sure the bottom of the 1 × 2 is flush with the bottom of the plywood. Cut two plywood shelf supports (M) to size, nail them to the shelf, then nail the whole shelf unit to the rest of the tool rack.

5 Sand all edges smooth. Cover with wood sealer. Secure the tool rack to the wall.

Materials: ¾-inch plywood, 1 × 4 and 1 × 2 lumber, perforated hardboard, glue, finish nails, screws, and wood sealer.

KID-SIZED PICKUP

Your toddler will spend many happy hours "behind the wheel" of this imaginative vehicle. Made from plywood and trimmed with 1 × 1s and 1 × 2s, it rolls around easily on two-inch rubber casters. And you just can't beat the mileage.

1 Before beginning construction, study the drawing and accompanying dimensions.

2 Begin by cutting out all members to size. Be sure to double check your measurements before cutting.

3 Begin assembly by sandwiching the floorboard (A) between the two body side panels (B). Glue and nail together. Secure the front panel (C) as shown.

4 Next secure diagonal panel (D) between the side panels, and then the dashboard (E).

5 Assemble the seat assembly (F,G), then glue and nail it to the floorboard.

6 Build truck box assembly, butting members H, I, J, and K as shown. Cut a rear box support (L) to size, secure it to the floorboard, and glue and nail the box assembly to the body (see sketch).

7 Secure the grill (M), headlights (N), spacer (O), and bumper (P) to the front panel, then mount the "wheels" (Q) and spacers (R,S) as shown.

8 Hang the door (T), using T-hinges and a wooden latch. Glue and nail taillights (U) into place, and apply 1 × 1 trim as shown. Complete assembly by screwing casters to the underside of the floorboard.

9 Fill all voids with wood putty, then sand smooth. Finish the truck as desired. Here, high-gloss latex paint was used.

Materials: ¾- and ⅜-inch plywood, 1 × 2 and 1 × 1 lumber, wood glue, finish nails, hinges and screws, rubber plate casters, wood putty, and finish.

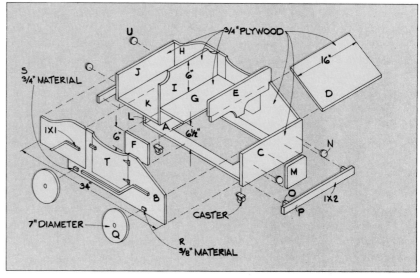

SLIM, TRIM SHELF UNIT

You don't need much floor space for this space-efficient unit. It's a bit more than a foot deep yet chock-full of storage.

To duplicate this unit's jaunty look, combine stained areas with brightly painted doors and drawer fronts.

1 Plan the dimensions before beginning construction. The unit shown is 7 feet tall and about 4 feet wide. Tailor yours to the available space.

2 From ¾-inch plywood, cut three uprights (A) and 14 shelves (B). Using 1 × 2 ledgers for support, glue and nail the upper two pair of shelves between the uprights. Cut two 1 × 4 stretchers (C) the same length as B and four 1 × 4 trim pieces (D). Glue and nail the trim pieces to the longer 1 × 4s, then glue and nail the assembly between the uprights, 3 inches in from the front edge.

3 For each upper cabinet, position and secure a shelf for the top and bottom, again using 1 × 2 ledgers; then cut a vertical divider (E) notching the front edge of it to receive the 1 × 2 frame pieces (F,G). Cut frame pieces (F,G) to size. Glue and nail the frame pieces together, nail the frame to the divider, and slip the whole assembly into the cavity. Nail into place. Cut two plywood doors (H), and hinge them to the 1 × 2 frame.

4 To create the lower cavities, simply secure another pair of shelves (B) to the uprights.

5 For each center drawer assembly, cut a shelf (I) and two side pieces (J) from ¾-inch plywood. Glue and nail together as shown, and secure them to the inside of the uprights. Cut the drawer members, using ¾-inch plywood for the front (K), ½-inch plywood for the sides (L) and back (M), and ¼-inch plywood for the bottom (N). Allow for the drawer guides when cutting. Cut

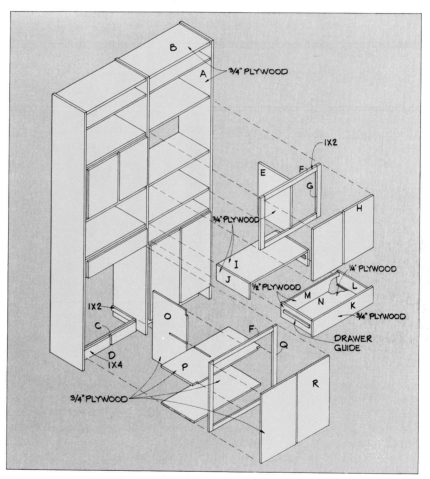

a ¼ × ¼-inch dado in the front, side, and back pieces to receive the bottom. Glue and nail together. Attach drawer guides to the drawer and to the surround, and slide the drawer into place.

6 For each base cabinet, position and secure the remaining two shelves. Then cut a vertical divider (O) to fit between them. Notch the divider to accept the 1 × 2 frame and the shelf (P) to be positioned later. Nail the divider into position, then cut the

shelf (P). Notch it as shown, then nail it into place. Frame the cabinet with 1 × 2s (F,Q). Hinge ¾-inch plywood doors (R).

7 Fill voids with wood putty, then sand the unit and finish as desired. Add catch and pull hardware.

Materials: ¾-, ½-, and ¼-inch plywood; 1 × 4 and 1 × 2 lumber; pivot hinges; catch and pull hardware; glue; nails; wood putty; drawer guides; and finish.

HEAVY-DUTY STORAGE CENTER

Storing your coffee-table books, record albums, and the like calls for more support than your ordinary storage unit provides. With this contemporary version, the "fins" are tied to the wall studs with wood screws. So go ahead and pile it on; this tough customer can take it.

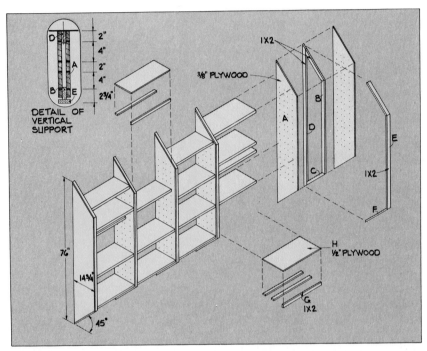

1 This project occupies about 5½ feet of wall space; the uprights are 76 inches tall. Adjust these dimensions, if desired.

2 Start by drawing a pattern for the upright (A) on a sheet of ⅜-inch plywood. Cut out this piece, then cut enough more so you'll have two for each upright. Sandwich all but two of the members, mark the position of the shelf support holes, and drill ⅜-inch holes.

3 Cut 1 × 2 upright spacer strips (B,C) to size, mitering the ends. Construct the uprights as shown using glue and nails. Make sure to align the dowel holes.

4 Locate the wall studs, then secure a 1 × 2 spacer (D) to them as shown using 3½-inch wood screws. Make certain the spacers are parallel to each other.

5 To tie the uprights to the wall, lift them up, slip them onto the spacers (D), and fasten with wood screws. Face the uprights with 1 × 2s (E,F), miter the ends as needed and nail into place.

6 Construct the shelves as shown, using 1 × 2s (G) and ½-inch plywood (H). These shelves vary in width—7, 10, and 14 inches. Cut the needed number of 6-inch-long ⅜-inch dowels. Set them aside.

7 Sink nails and fill all nail holes, then sand the shelves and uprights smooth. Finish as desired. Lift shelves into position and secure with dowels.

Materials: ⅜- and ½-inch plywood, 1 × 2 lumber, ⅜-inch wood dowel, glue, wood screws, finish nails, wood putty, and finish.

CHIC SIDEBOARD/ DIVIDER

With this unit between your kitchen and dining areas, you create not only a handy work surface and storage center, but **also a see-through divider that separates yet visually unifies the two areas. It's inexpensive to build, too.**

1 Carefully plan the dimensions of the unit. The one shown measures 18 × 36 × 84 inches.

2 Cut two uprights (A) from ¾-inch plywood. Using a 9-inch radius, round the top end of each upright. Clamp the two panels together and drill holes as shown to receive the two 1½-inch-diameter wood dowels (C). Cut two 1 × 3s (B) to serve as tray rack rails, and 17 lengths of ¼-inch dowel (D). Drill 1 × 3s to receive dowels set on 2-inch centers, and secure dowels to the 1 × 3s using wood glue.

3 For the lower portion of the unit, cut rectangular shelves (E) from ¾-inch plywood. Cut another shelf (F) ¼ inch narrower than the two other shelves. Cut a ⅜ × ¼-inch-deep rabbet in the back edge of the two deeper shelves (E) for the back.

4 Cut the ¼-inch hardboard back (G) to size. Set it aside. Cut two lengths of metal sliding door track, then screw them to the shelves where shown.

5 To assemble the sideboard, first position the shelves between the uprights and nail them into place. Then, nail the cabinet back to the shelves as shown. Nail and glue the plate rack into place (see sketch), and cut the 1½-inch dowels so they extend 3 inches beyond the outside edge of each of the uprights. Position and secure the dowels.

6 Cut hardboard doors (H) to size; drill holes for the pulls.

7 Fill all voids with wood putty, then sand the unit smooth. Finish as desired. When the finish dries, snap finger pulls into the door

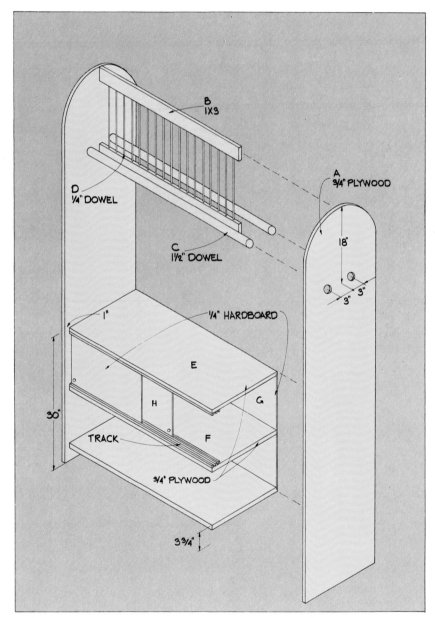

panels, then insert the doors into the track.

Materials: ¾-inch interior plywood, ¼-inch tempered hardboard, 1 × 3 lumber, 1½- and ¼-inch wood dowel, sliding door track, snap-in finger pulls, wood glue, wood putty, finish nails, and desired finish.

"ANTIQUE" BOOKCASE AND DIVIDER

Turned spindles, cabinet overlay molding, and antique finish give this ceiling-high unit old-world flavor and charm. Storage space on the shelves above and in the cabinet below make this more than just a good-looking divider.

1 When planning the dimensions of your unit, figure on the cabinet's being about 3 feet tall.

2 Cut the cabinet's sides (A), bottom (B), and shelves (C) to size. Then assemble members.

3 Cut the four shelves (D) to size (they should overlap the cabinet 1½ inches in front and ½ inch on either side). Shape the edges of each shelf with a router. Nail the bottom shelf to the cabinet. Position and secure the spindles and remaining shelves (see detail). Trim top shelf with molding (E,F).

4 Cut cabinet door (G) to size, glue and nail the moldings (H,I,J) to it, and hinge the door. Add knob and latch hardware.

5 Fill voids with wood putty, sand unit, and apply finish.

Materials: ¾-inch plywood, spindles, molding, glue, nails, hinges, wood putty, and finish.

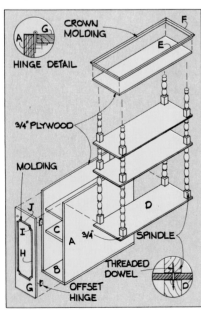

STYLISH OAK END TABLE

Warm oak lumber contrasted against crisp white plastic laminate provides the textural interest of this end table. Casters provide mobility, and storage areas beneath the tabletop make ideal nooks for just about anything.

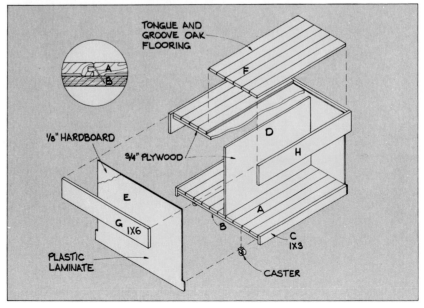

1 Plan the end table's dimensions carefully before beginning construction. The table shown is 22 inches tall.

2 Start by butting together the edges of several pieces of oak flooring (A). Cut the flooring to the desired length. Cut two ¾-inch plywood subplatforms (B) to the same dimensions as the flooring. Set one aside. Nail and glue the flooring to the plywood (see nailing detail).

3 Cut 1 × 3s (C) to size, then nail and glue them to the base platform (A,B).

4 Cut a ¾-inch plywood divider (D) to size, then cover each side with plastic laminate (see page 93). Glue and nail the divider between the top and bottom plywood subplatforms, making sure everything is square.

5 Cut two ⅛-inch hardboard end panels (E) to size. Nail the panels to the framework. Then apply laminate to both sides of panels.

6 Cut several more lengths of oak flooring (F) to cover the top subplatform. Glue and nail to the plywood.

7 Trim out the top of the end table with 1 × 6s (G,H). Either butt the members or, for a more finished look, miter the corners.

8 Fill all nail holes with wood putty, then sand all surfaces smooth and finish. Fasten casters to the underside of the unit.

Materials: ¾-inch plywood, ⅛-inch hardboard, 1 × 6 and 1 × 3 oak, tongue-and-groove flooring, plastic laminate and contact cement, glue, finish nails, wood putty, casters, and finish.

ZIGZAG BOOKSHELF

This bookshelf will add storage space as well as a unique touch to any room. Half-lapped frames hinged together in accordion fashion provide plenty of support for the 6-foot-long shelves that thread their way through the uprights.

1 The unit pictured here is 72 inches tall; each panel is 18 inches wide. You may alter these dimensions somewhat, but keep the shelves short enough so they don't cantilever too far beyond the end supports. We used good-quality 1 × 3 and 1 × 12 lumber, but ¾-inch plywood would work equally well.

2 From 1 × 3 lumber, cut eight uprights (A). Cut a 2½ × ⅜-inch notch in each end of each upright. Then carefully mark the location of the crossmember shelf supports (B) on the uprights and cut 2½ × ⅜-inch dadoes to accommodate the crossmembers.

3 Cut 24 crossmembers (B) to size, and notch each end of each member as before. Construct the frames using glue and finish nails. Miter the edges of the uprights at 45 degrees as needed.

4 Hinge the panels together as shown in the detail. Cut four appropriately sized shelves (C) from 1 × 12 lumber.

5 Fill all voids with wood putty and sand everything smooth; finish as desired. Here, two coats of varnish were applied. When finish dries, slide shelves into position.

Materials: 1 × 12 and 1 × 3 lumber, butt hinges, glue, finish nails, wood putty, and finish.

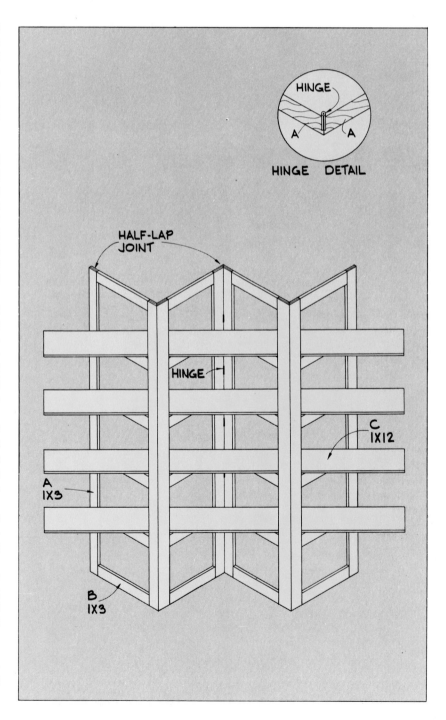

GO-ANYWHERE PARTY CENTER

With this versatile caddy, serving food and beverages at backyard parties is a breeze. Just equip it ahead of time and roll it to a convenient spot for serving. The cart stores everything from wine bottles to utensils, and features a durable, two-level work surface.

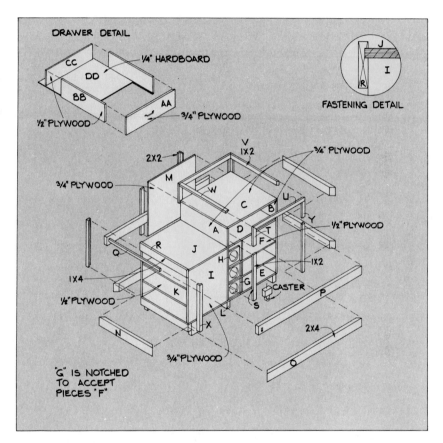

1 The unit shown measures 27 × 53 × 38 inches. The lower work surface is 30 inches high. Plan the dimensions carefully before beginning. The size of ceramic tiles you use will determine the dimensions of the work surface. Use materials suitable for outdoor use.

2 Start by cutting members A,B,C,D, and E from ¾-inch exterior plywood. Glue and nail them together as shown. Then cut ½-inch plywood shelves (F) and a vertical divider (G), notching them as shown. Cut the beverage nook facers with 4¼-inch cutouts (H) to size; glue and nail F,G, and H into place.

3 Construct the frame for the shorter end of the unit by gluing and nailing members I and J as shown. Add shelves (K). Glue and nail 1½ × 2¾-inch blocking (L) to the unit's base (see sketch). Also mount scrap 2 × 4 legs and casters at this time. Glue and nail member M into place as shown to close up the back of the drawer cavity.

4 Fill all voids with wood putty, then apply two or more coats of exterior paint to all surfaces.

5 Cover each of the work surfaces and the vertical divider with ceramic tile.

6 Using rough-sawn cedar 2 × 4s, cut and trim pieces N,O, and P to size, then secure them to the cart with lag screws (or galvanized nails) and glue. Be sure to make a ¾ × 1½-inch opening in each P member to accept a 1 × 2 handle (Q). Position and secure the handle.

7 Secure cedar trim members R,S,T,U,V, and W to the cart as shown with glue and nails. Then cut a ¾ × ¾-inch rabbet from a long length of cedar 2 × 2. From this piece, cut members X and Y. Glue and nail these into place.

8 Build the drawer as shown, using ¾- and ½-inch plywood and ¼-inch hardboard for the drawer members (AA, BB, CC, and DD). Fill nail holes and sand drawer smooth; finish with exterior paint. Attach a gate pull to the drawer front.

Materials: ¾- and ½-inch exterior plywood; ¼-inch hardboard; 2 × 4, 2 × 2, 1 × 4, and 1 × 2 lumber; waterproof glue; lag screws (optional); finish nails; wood putty; ceramic tiles, adhesive, and grout; plate-type casters; gate pull; and the desired water-resistant finish.

GRAND-STYLE WINE CLOSET

Whether you're a wine connoisseur or a beginning fancier, you'll appreciate this unit. Roomy enough for all of your favorites, it also provides easy access for selection.

1 Plan the dimensions of your unit before beginning. This unit measures 16 × 42 inches.

2 From ¾-inch plywood, cut two uprights (A) just shy of floor-to-ceiling height, and the top and bottom pieces (B). Glue and nail the pieces together.

3 Cut a ¾-inch plywood vertical divider (C) and two horizontal dividers (D) to size. Center the vertical divider in the frame (A,B), then glue and nail it in position. Nail 1 × 2 ledgers (E) to the uprights and vertical divider. Then position the horizontal dividers and nail them to the ledgers.

4 For each of the longer diagonal dividers (F), hold a 1 × 10 up to the opening, mark it, and cut the piece to fit. Position and secure by toenailing in place. Follow a similar procedure for the shorter diagonal dividers (G).

5 For the doors, cut four 1 × 4 uprights (H) to match the height of the unit and six 1 × 4 crosspieces (I). Glue 1 × 4s together to form the frame; secure each joint with corrugated fasteners. Or fasten using half-lap joints, glue, and nails. Cut four pieces of 1 × 8 (J) of the shape shown to trim the top and bottom. Glue and nail them to the door frame. Cut lengths of chicken wire to fit on the back side of the doors and staple into place. Hinge the doors to the closet;

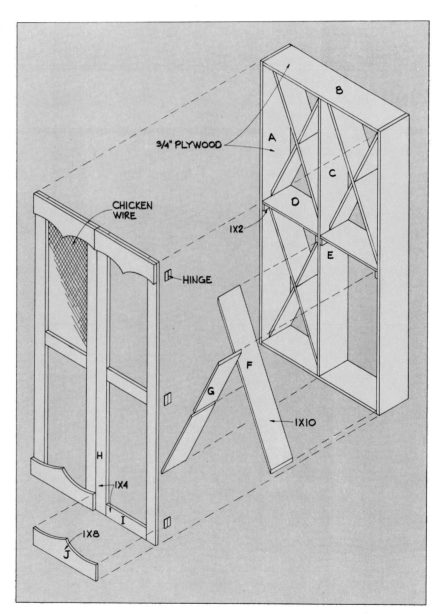

add pulls and catch hardware.

6 Fill all voids with wood putty, sand smooth, and finish.

Materials: ¾-inch plywood; 1 × 10, 1 × 8, 1 × 4, and 1 × 2 lumber; hinges; pulls; catch hardware; corrugated fasteners; glue; screws and decorative washers; finish nails; wood putty; chicken wire mesh; and desired finish.

CONTEMPORARY WOOD AND GLASS DINING TABLE

A striking departure from the traditional, this innovative dining table design combines the warmth of wood with glass in truly dramatic fashion. Exterior cedar siding covers the table's supports; cedar 1 × 3s trim the glass tabletop.

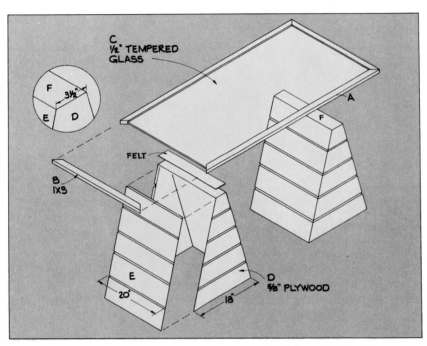

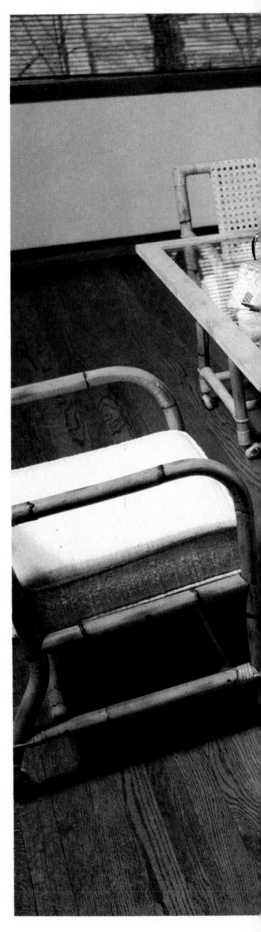

1 The unit shown here measures 40 × 66 × 30 inches. You can increase the length of your table, if desired, but doing so may make it necessary to add another table support.

2 To construct the tabletop, cut full-dimension 1 × 3 frame members (A,B) to size. Cut a ½ × ½-inch groove in the inside edge of each member, then miter the ends of each. Fit the frame members around the ½-inch tempered glass top (C), and glue and nail them together.

3 For each table support, cut two end panels (D), two side panels (E), and one top panel (F) from ⅝-inch exterior plywood siding. The end panels taper from 18 inches at the base to 3½ inches at the top. All of the panels are mitered for a snug,

finished appearance. The supports at the base measure 18 × 20 inches.

4 Assemble the table supports using wood glue and finish nails. Then top each support with a piece of felt material to protect the glass tabletop.

5 Sand the table frame members smooth, rounding the corners and edges. Sink nails and fill all voids with wood putty, then finish the unit as desired. If desired, you can leave the wood its natural color, but you should apply a coat or two of polyurethane varnish nonetheless.

Materials: ⅝-inch exterior plywood siding, full-dimension 1 × 3 lumber, ½-inch tempered glass, wood glue, finish nails, wood putty, and desired finish.

SPACE-SAVER SEWING CENTER AND CABINET

With this swing-out unit, you don't need lots of extra space at your house for a sewing center. When the sewing surface is tucked away in its storage nook, the unit protrudes only about 18 inches from the wall.

1 Carefully plan the dimensions of each module before beginning construction. The unit shown stands 84 inches tall, and the roll-out sewing cart is 30 inches tall.

2 For each end module, cut two notched uprights (A) from ¾-inch plywood. Cut grooves in them as shown to receive adjustable shelf standards. Secure the standards to the uprights, then cut a plywood crossmember (B) to size. Cut a ⅜ × ¼-inch-deep rabbet in all three members to accept the hardboard back.

3 Cut a ¼-inch hardboard back (C) to size, then glue and nail A,B, and C together. Cut three 1 × 4 stretchers (D) to further stabilize the module. Nail and glue into place.

4 Cut shelves (E,F,G) and a cabinet door (H) to size, then position and secure them to complete the module.

5 For the center unit, cut two uprights (I) and two crossmembers (J,K). Cut grooves in the uprights for the shelf standards and rabbet all four members to accept a ¼-inch hardboard back (L). Cut the back, then assemble the frame with glue and nails. Cut a shelf (M), a 1 × 4 facer (N), and two doors (O) to size. Position and secure them to complete this module.

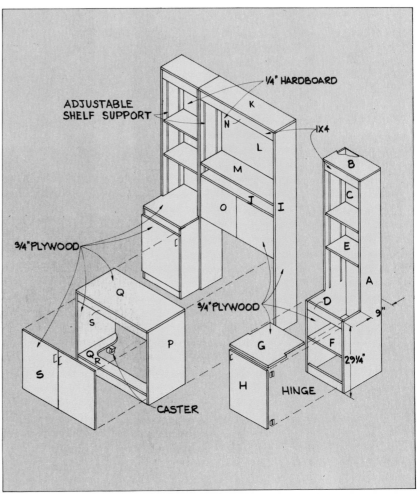

6 For the sewing cart, cut uprights (P), a top (Q), and a sculptured footrest (R) from ¾-inch plywood (don't forget to allow for the casters in your measurements). Also cut two 1 × 4 facers (S), then glue and nail the members together as shown. To simulate cabinet doors on the cart, cut a plywood panel (T) to size. Dado as shown. Glue and nail the panel into place.

7 Sink all nails, fill voids with wood putty, sand the unit smooth, and finish as desired. Add catch and pull hardware. If desired, bolt the modules together. Fasten casters to the underside of the sewing cart.

Materials: ¾-inch plywood, ¼-inch hardboard, 1 × 4 lumber, adjustable shelf standards and clips, hinges, catch and pull hardware, casters, glue, nails, wood putty, and finish.

MULTIDUTY WORKTABLE

Whether you're drafting house plans or jotting down a quick note to friends, this table is a convenient place to do it. The tapered top surface lifts off easily for those situations that require a desk-height table.

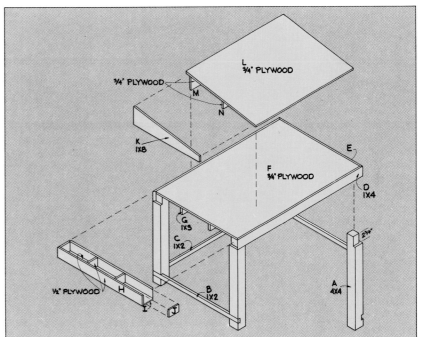

1 Begin by planning the dimensions of your table. The one shown measures 38 × 62 inches. The lower table surface is 30 inches high; the tapered upper one, about 36 inches.

2 Cut four 4 × 4 legs (A) to size. Notch them as shown. Note that the notches at the top of the legs are 2¾ inches wide; this enables 1 × 4 members D,E to cover the raw plywood edges of member F.

3 Cut lower rails (B,C) and upper rails (D,E) to size, then assemble the table's frame, except the left-hand E member, using glue and nails.

4 Cut a plywood top (F) to fit within the frame, and two 1 × 3 tabletop braces (G). Glue and nail to each other, then to the table frame.

5 For the side tray, cut members H,I, and J to size, glue and nail them to the remaining table frame member (E), and secure the whole assembly to the table.

6 For the upper table surface, cut the tapered side members (K) and the work surface (L) to size, then glue and nail them together. Strengthen the surface with ¾-inch plywood support strips (M,N).

7 Fill all voids with wood putty, sand all surfaces smooth, then finish the table as desired. Glue a drafting table vinyl material (available at drafting supply stores) to both tabletops.

Materials: ¾- and ½-inch plywood; 4 × 4, 1 × 8, 1 × 4, 1 × 3 and 1 × 2 lumber; drafting table vinyl and adhesive; glue; wood putty; nails; and finish.

PLANNING AND BUILDING BASICS

Have you ever heard the saying, "It's not hard once you know how to do it"? Well it's true, especially when it comes to building projects for your home. Once you've mastered the basics, each project you do becomes a series of accomplishable steps rather than a major undertaking.

So, do yourself a favor and spend a few minutes learning about the steppingstones to professional-looking projects. It'll be worth your while.

COMMON CONSTRUCTION MATERIALS

The materials you use for construction will vary, depending on the item's intended use. So when making your selection, ask yourself these questions: Are you constructing something for indoor or outdoor use? Is the item strictly utilitarian, or will it be suitable for use in a living room? Is it intended for light-duty use, or will it be a long-lived project subject to considerable use—and abuse?

Hardboard

Hardboard is available in 4x8-foot sheets and comes in ⅛- and ¼-inch thicknesses. Standard hardboard is an excellent choice for cabinetwork, drawer bottoms, and concealed panels.

You can also get hardboard perforated with holes spaced about one inch apart. Perforated hardboard is recommended for building storage for soiled laundry and for the backs of hi-fi cabinets. The quarter- and eighth-inch perforated hardboard lends itself to storing garden equipment and tools, too, as its holes accept hooks designed for this purpose. To expand or change the arrangement, just switch the hooks around. If the project will be subject to dampness, use tempered hardboard.

Particle board, chip board, and flake board, also members of the hardboard family, have a coarser grain structure, are lighter in color, and are available in thicknesses up to ¾ inch. These products are made of granulated or shredded wood particles forced together under pressure with a binder at high temperatures.

Plywood

Plywood also comes in 4x8-foot sheets, though larger sheets are available on special order. Thicknesses range from ⅛-inch to ¾-inch. For light-duty storage, the ¼- and ½-inch thicknesses are adequate. If you are planning to build an outdoor storage unit, specify *exterior grade* when making your purchase. Exterior grade plywood has its layers glued together with a waterproof glue to withstand rain.

The surfaces of plywood sheets are graded A, B, C, and D—with A the smoother, better surface and D the least desirable appearance. Choose AA (top grade, both sides) only for projects where both sides will be exposed; use a less expensive combination for others.

Solid Wood

Plain, ordinary wood still ranks as the most popular building material. Wood is sold by the "board foot" (1x12x12 inches). One board foot equals the surface area of one square foot, with a nominal thickness of one inch.

Wood is marketed by "grade." For most building projects No. 2 grade will satisfy your needs. This grade may have some blemishes, such as loose knots, but these don't reduce the strength of the wood.

If you're planning to build a unit that will be part of a room's decor, you should buy *select lumber*—a grade that's relatively free of blemishes.

Remember, too, that outdoor projects are a different subject.

Redwood or cedar is preferable, but if you use a soft wood, be sure to treat it for moisture resistance. Or buy pressure-treated wood.

You can buy boards up to 16 feet in length and 12 inches in width, though occasionally a lumberyard may have somewhat wider or longer boards.

Wood is divided into two categories. Softwoods, used commonly for general construction, come from trees that don't shed their leaves in the winter: hemlock, fir, pine, spruce, and similar evergreen cone-bearing trees. Hardwoods come from trees that do shed their leaves: maple, oak, birch, mahogany, walnut, and other broad-leaved varieties.

All lumber is sold by a nominal size. A 2x4, for example, does not measure two by four inches. It's actually 1½x3½ inches (though the nominal *length* of a 2x4 is usually its true length). The drawing shows nominal sizes, as well as the actual sizes, of most pieces of common lumber.

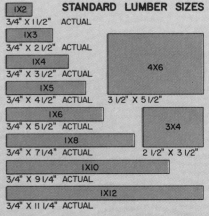

STANDARD LUMBER SIZES

1X2	3/4" X 1 1/2" ACTUAL
1X3	3/4" X 2 1/2" ACTUAL
1X4	3/4" X 3 1/2" ACTUAL
1X5	3/4" X 4 1/2" ACTUAL
1X6	3/4" X 5 1/2" ACTUAL
1X8	3/4" X 7 1/4" ACTUAL
1X10	3/4" X 9 1/4" ACTUAL
1X12	3/4" X 11 1/4" ACTUAL
4X6	3 1/2" X 5 1/2"
3X4	2 1/2" X 3 1/2"

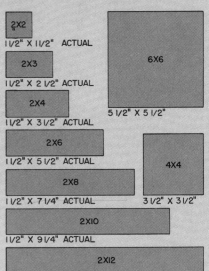

2X2	1 1/2" X 1 1/2" ACTUAL
2X3	1 1/2" X 2 1/2" ACTUAL
2X4	1 1/2" X 3 1/2" ACTUAL
2X6	1 1/2" X 5 1/2" ACTUAL
2X8	1 1/2" X 7 1/4" ACTUAL
2X10	1 1/2" X 9 1/4" ACTUAL
2X12	1 1/2" X 11 1/4" ACTUAL
6X6	5 1/2" X 5 1/2"
4X4	3 1/2" X 3 1/2"

SUPPORT SYSTEMS

Any item you construct, no matter how light, must be capable of supporting itself as well as its "payload". Even a simple box has a support system: its sides are self-supporting, each one serving to support and strengthen its neighbor.

How to Attach Things to Walls

Many items, such as shelves and wall-hung cabinets, depend on the wall as part of their support system. However, you can't always drive a nail or insert a screw just anywhere in a wall. For best stability, drive them into the studs of the wall.

Locating studs. One way of locating wall studs is to rap the wall with your knuckles. Listen for a "solid" sound. (Thumps between the studs will sound hollow.) This works fine if you have excellent hearing.

A far easier way is to buy an inexpensive stud finder. Its magnetic needle will respond to hidden nails, indicating the presence of a stud.

Locating one stud does not necessarily mean that the next stud is 16 inches away, though. It should be, but many times it isn't. For example, if the framework of a door or wall falls 20 inches away from the last stud, the builder may have left a 20-inch gap between them. Or, a stud may have been placed midway, leaving 10-inch spaces on either side.

Fastening to hollow-core walls. Quite often, because of physical requirements, you will need to make an installation between studs into a hollow plaster wall.

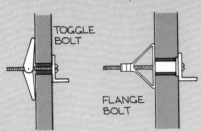

TOGGLE BOLT

FLANGE BOLT

What then? The answer is to use flange or toggle bolts. They distribute their load over a wide area,

and if used in sufficient number and with discretion, they'll hold a fairly heavy load.

Fastening to masonry. Attaching items to a masonry wall is not difficult. if you're working with a brick, concrete, or cinder block wall, use a carbide-tipped drill to make a hole *in the mortar*. Make the hole deep and wide enough to accept a wall plug. Then insert the screw or bolt to fasten the item in place (see sketch).

Another method of fastening to

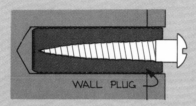

WALL PLUG

masonry walls is to drill a ½-inch hole in the *mortar* and pound a hardwood dowel into the hole. Bevel the end of the dowel and lightly coat it with grease before driving it in place. Then drill a pilot hole in the middle of the dowel and continue with the fastening.

If by chance you must drill into the brick part of a wall rather than the mortar, don't despair. Again use a carbide-tipped drill, but this time start with a ¼-inch bit, and finish with the larger size desired.

How to Mount Units On a Base

If your project is any type of cabinet, a base is a good idea. A base should provide toe space of at least 3½ inches in height and 2¾ inches in depth. If you plan to mount the unit on casters, you'll automatically get toe space that makes the project convenient.

Box base. This easy-to-build recessed base consists of a four-sided open box installed at the

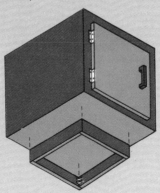

bottom of the cabinet or storage unit. Since appearance is not a factor, you can construct the box with simple butt joints and secure it to the cabinet with steel angle brackets installed along the inside of the base (see sketch).

Leg base. Four short, stubby legs also make a good base. Commercial legs come with their own mounting plate, which is screwed to the bottom of the cabinet before the leg is screwed into place (see sketch). You can also install home-built legs with hanger bolts.

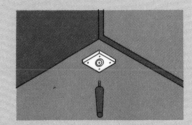

These bolts have a "wood" thread on one end and a coarse "machine" thread on the other end. Drill an undersize hole in the cabinet for the machine end, insert the hanger bolt using pliers and screw the leg into place.

A good source for low-priced legs is a lumberyard that does millwork. Quite often, they'll have a bin full of legs of all sizes that may have slight imperfections or chips which won't affect their serviceability.

How to Mount Shelves

Shelves are a quick and easy way of getting additional storage space in your home, shop, or garage. The best material for shelving is ¾-inch plywood or pine boards—8, 10, 12 inches wide, depending on the items to be stored. To prevent sagging, install a shelf support every 30 inches. And don't use hardboard or chip board, as they tend to bow under heavy loads.

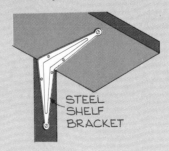

STEEL SHELF BRACKET

Shelf brackets. The easiest way to mount a shelf is by means of steel shelf brackets sold in hardware stores (see sketch). Ask for brackets whose short leg is nearly equal to the *width* of the shelf you plan to install. And always mount the brackets with the *long* leg against the wall. Screw the brackets into the wall and space them about 30 inches apart. For heavy loads, shop around for brackets that have gussets connecting the two legs. Brackets without gussets tend to sway under heavy loads.

Cleats and angle brackets. The narrow space between two walls is an ideal location for shelving. Simply install a pair of cleats at the heights where you want shelves (see sketch). Use cleats that are at least ¾ inch thick and as long as the shelf is wide.

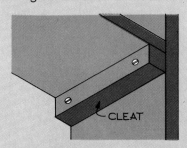

CLEAT

If the walls are of masonry, secure the cleats with so-called steel cut nails (wear goggles when driving these, as they may break off if not struck head-on). Secure the cleats with screws if the walls are of wood, or use flange bolts if they're hollow.

You can also use small steel angle brackets. Mount two under each side of the shelf as shown.

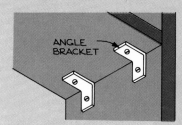

ANGLE BRACKET

Dowels. Another method of supporting shelves is with dowels. Drill holes equal to the diameter of the dowels, and bore them deep enough to accept at least ½-inch of dowel length. (Make sure both left and right holes are the same height; you might use a level on the shelves to ensure exact mounting.)

Use ¼-inch dowels for light-duty shelves and ⅜-inch dowels for shelves supporting heavy loads. Beveling the dowel ends

will make them easier to insert into the holes. To change shelf spacing, simply drill additional holes.

Dado cuts. This method of supporting shelves has long been a favorite with master cabinetmakers. First, determine the height of the shelf, mark the uprights, and make your cuts. Then cut the shelf to fit.

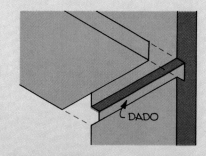

DADO

Metal tracks and brackets. You can recess or surface-mount these handy shelf supports. Shelf brackets, specially designed to fit into the track slots, are made to accept 8-, 10-, and 12-inch-wide shelves. Special brackets which adjust to hold shelves at a downward slope also are available and are used to hold dictionaries and reference books.

These tracks and brackets are available in finishes to match the decor of practically any room.

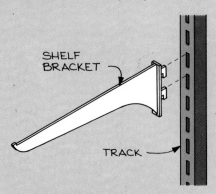

SHELF BRACKET

TRACK

When installing shelves in a cabinet, mount two tracks on each side of the cabinet and use small clips to hold the shelves in place. To change the spacing between shelves, just remove the clips and reposition.

Furring strips. These are especially useful for supporting and erecting shelves in the garage or workshop. Use 2x4s bolted or screwed to the wall and short lengths of 1x4s for shelf supports, as indicated in the drawing. Note that one end is dadoed into the 2x4 (½-inch depth is enough). The

front end of the shelf support bracket is supported by a 1x4 cut at a 45 degree angle at the bottom and engages a cutout called a *bird's mouth* at the top. Toenail

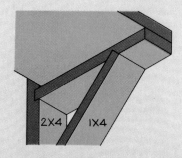

2X4   1X4

the lower end of the 1x4 into the 2x4. There's no need to nail the upper end, as the weight of the shelf will keep it in place.

Support from above. While most shelves are supported from the bottom, you can also support them from the top. This top support method is especially applicable in basement areas where the joists are exposed. You can nail 2x4s to the joists and fit any type of

project—open shelves, a cabinet, even a work surface between them. If the project to be suspended will run perpendicular to the joists, be careful to plan the length so that it will match the spacing of the joists.

Another way to support shelving from the top is use threaded rods (see sketch above). Choose rods from ¼- to ¾-inch diameter according to the load you'll support. Drill holes in the shelves slightly oversize. To attach the upper end of the rod, drill holes in 2x2 scraps and screw to the joists. Insert the rod and add a nut and washer to the top.

Then install the shelves with a nut and washer on both top and bottom. Tighten the nuts securely to give the shelves as much stability as possible.

WOOD JOINERY TECHNIQUES

No matter what material you're planning to use, it will have to be cut to size—measure twice and cut once is a good rule— then put together using glue, nails or screws, and one of these joints.

Butt Joints

The simplest joint of all, the butt joint, consists of two pieces of wood meeting at a right angle and

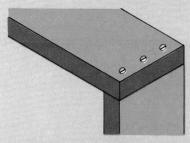

held together with nails, or preferably, screws (see sketch). A dab of glue before using the nails or screws will make the joint even more secure. But don't use glue if you're planning to take the work apart sometime later.

When reinforced by one of the six methods illustrated, the butt joint is effective for making corner

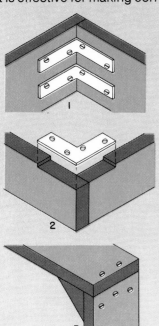

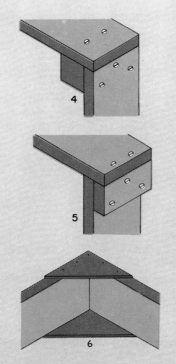

joints. Two common fasteners are corner braces (1), and flat corner plates (2). Using scrap wood, you can reinforce the joint with a triangular wedge (3), or with a square block (4). A variation of the square block places the block on the outside of the joint (5). Finally, a triangular gusset made from plywood or hardboard will also serve to reinforce a corner butt joint (6).

When a butt joint is in the form of a T—for example, in making a framework for light plywood or hardboard—you can reinforce it with a corner brace, T plate, or corrugated fasteners.

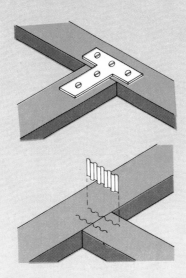

For really rough work, you can drive in a couple of nails at an

angle, or toenail (see sketch). A variation of this is to place a block of wood alongside the crosspiece

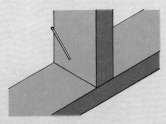

and secure it with a couple of nails.

A close cousin to the T joint and the butt joint is the plain overlap joint. It is held in place with at least two screws (see sketch). For extra reinforcement, apply glue between the pieces of wood.

Butt joints are an excellent means of securing backs to various units, especially when appearance is not a factor. Simply cut the back to the outside di-

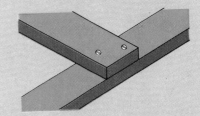

mensions of the work, then nail in place . . . it's called a flush back.

Lap Joints

On those projects where appearance is vital, consider full and half-lap joints. To make a full lap joint, cut a recess in one of the pieces of wood equal in depth to the thickness of the crossmember (see sketch).

The half-lap joint is similar to the full lap joint when finished, but the technique is different. First, cut a recess equal to half the

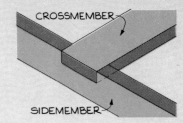

CROSSMEMBER

SIDEMEMBER

thickness of the crossmember halfway through the crossrail. Then, make a similar cut in the opposite half of the other piece (see sketch on the next page).

Butt joints and overlap joints do

not require any extra work besides cutting the pieces to size. However, full and half-lap joints

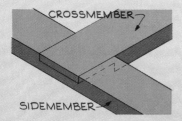

require the use of a backsaw and a chisel. For a full-lap joint, mark off the thickness and width of the crossmember on the work in which it is to fit.

Use the backsaw to make a cut at each end that's equal to the thickness of the crossmember, then use a chisel to remove the wood between the backsaw cuts. Check for sufficient depth and finish off with a fine rasp or sandpaper. Apply white glue to the mating surfaces and insert two screws to hold the joint securely.

Dado Joints

The dado joint is a simple way of suspending a shelf from its side supports. To make a dado joint, draw two parallel lines with a knife

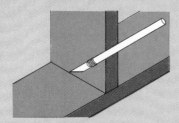

across the face of the work equal to the thickness of the wood it is to engage (see sketch). The depth should be about one-third of the thickness of the wood.

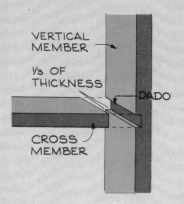

Next, make cuts on these lines and one or more between the lines

with a backsaw. Then, chisel out the wood to the correct depth.

You can speed the job immeasurably by using a router, a bench saw, or a radial arm saw. Any one of these power tools makes the cutting of dadoes an easy job — and provides much greater accuracy than can be achieved by hand.

If appearance is a factor, consider the stopped dado joint. In this type of joint, the dado (the cutaway part) extends only part way, and only a part of the shelf is cut away to match the non-cut part of the dado.

To make a stopped dado, first make your guide marks and chisel away a small area at the stopped end to allow for saw movement. Then make saw cuts

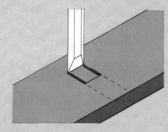

along your guide marks to the proper depth. Next chisel out the waste wood as shown in sketch.

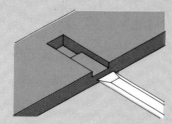

And finally, cut away a corner of the connecting board to accommodate the stopped dado.

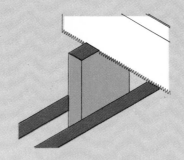

Rabbet Joints

The rabbet joint is really a partial dado. As you can see in the drawing at the top of the following column, only one of the meet-

ing members is cut away.

The rabbet joint is a simple one to construct, and it's quite strong, too. To ensure adequate strength, be sure to secure the meeting members with nails or screws and glue.

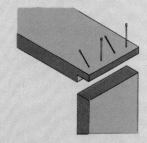

This joint is often used in the construction of inset backs for units such as cabinets and bookshelves (see the sketch below). To make this joint, rabbet each of the framing members, then care-

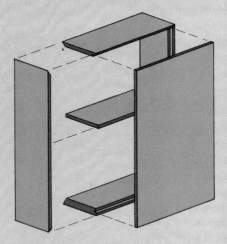

fully measure the distance between the rabbetted openings. Cut the back accordingly. Then use thin screws to secure the back to the unit.

Mortise and Tenon Joints

A particularly strong joint, the mortise and tenon joint is excellent when used for making T joints, right-angle joints, and for joints in the middle of rails. As its name indicates, this joint has two

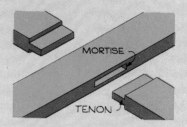

parts—the *mortise,* which is the open part of the joint, and the *tenon,* the part that fits into the mortise.

Make the mortise first, as it is much easier to fit the tenon to the mortise than the other way around. Divide the rail (the part to be mortised) into thirds and carefully mark off the depth and the width of the opening with a sharp pencil.

Next, use a chisel, equal to the width of the mortise, to remove the wood between the pencil marks. You can expedite this job by drilling a series of holes in the rail with an electric drill, a drill press, or even a hand drill. (If you have a drill press, you can purchase a special mortising bit that will drill square holes, believe it or not.) Mark the drill bit with a bit of tape to indicate the desired depth. Now use the chisel to remove the excess wood.

To make the tenon, divide the rail into thirds, mark the required depth, and use a backsaw to remove unwanted wood. If you have a bench or radial saw, the job of removing the wood will be much easier. Use a dado blade and set the blades high enough to remove the outer third of the wood. Reverse the work and remove the lower third, leaving the inner third intact.

To assemble, make a trial fit, and if all is well, apply some white glue to the tenon and insert it into the mortise. If by chance the tenon is too small for the mortise, simply insert hardwood wedges at top and bottom.

Use moderate clamping pressure on the joint until the glue dries overnight. Too much pressure will squeeze out the glue, actually weakening the joint.

Miter Joints

You can join two pieces of wood meeting at a right angle rather elegantly with a miter joint. And it's not a difficult joint to make. All you need is a miter box and a backsaw, or a power saw that you can adjust to cut at a 45 degree angle.

Since the simple miter joint is a surface joint with no shoulders for support, you must reinforce it. The easiest way to do this is with nails and glue (see sketch at the top of the following column). You'll notice that most picture

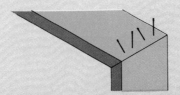

frames are made this way.

However, for cabinet and furniture work, you may use other means of reinforcement. One way is to use a hardwood spline as shown in the drawing. Apply glue to the spline and to the mitered

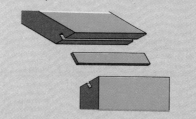

area and clamp as shown until the glue dries.

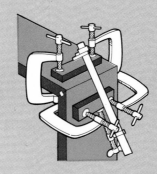

A variation of the long spline uses several short splines—at least three—inserted at opposing angles.

Dowels are a popular method of reinforcing a mitered joint, too. Careful drilling of the holes is necessary to make certain the dowel holes align. Use dowels that are slightly shorter than the holes they are to enter to allow for glue at the bottom. Score or roughen the

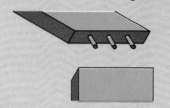

dowels to give the glue a better surface for a strong bond.

Dovetail Joints

The dovetail joint is a sign of good craftsmanship. It's a strong joint especially good for work subject

to heavy loads.

To make the joint, first draw the outline of the pin as shown and

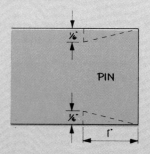

cut away the excess wood with a sharp backsaw. Place the pin over the second piece of wood and draw its outline with a sharp pencil. Make the two side cuts with the backsaw and an additional cut or two to facilitate the next step—chiseling away the excess wood. Then test for fit, apply glue and clamp the pieces until

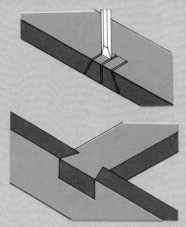

dry. This is the basic way to make most dovetail joints. However, it's much easier to make dovetail joints with a router and dovetail template, especially made for home craftsman use.

Corner Joints

These joints are used for attaching legs to corners for framing. A good technique for joining corners is the three-way joint involving a set of steel braces you can buy. First, insert the bolt into the inside corner of the leg. Then cut slots into the side members, and secure the brace with two screws at each end. Finally, tighten the wing nut.

A variation of the three-way joint uses dowels and a triangular ¾-inch-thick gusset plate for additional reinforcement. To make this joint, first glue the dowels in

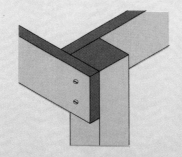

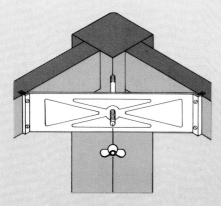

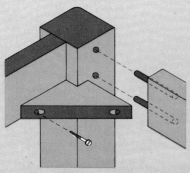

the vertical piece (see sketch). Let them dry completely, then finish the assembly.

A glued miter joint, reinforced with screws and glue, also makes a good corner joint. Make sure the screws do not penetrate the outside surface of the mitered joint.

Probably the strongest of the corner joints is the mortise and tenon (with mitered ends) reinforced with screws (see sketch). The miters on the ends of the tenons allow for a buildup of glue in the mortise, which in turn makes the joint stronger. Make sure that the holes you drill for the screws are not in line with each other.

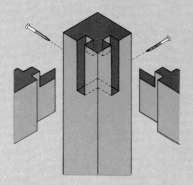

Otherwise, the wood may split. Use flathead screws and countersink the holes.

The simplest corner joint of all is a butt joint for the two horizontal members (see sketch). Instead

of being fastened to each other, the butted members are each

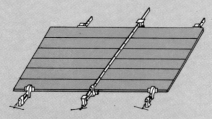

fastened to the corner post with screws.

Edge-to-Edge Joints

Whenever an extra-wide surface is required, such as a desk top, workbench, or a large storage cabinet, this joint fills the bill. To make it, glue together two or more boards, then hold securely with either bar or pipe clamps. If the boards have a pronounced grain, reverse them side-to-side

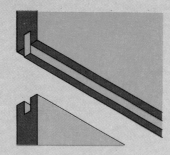

to minimize warping. For additional strength, screw cleats to the underside of the boards.

You also can use hardwood splines to join several boards. Cut a groove the exact width of the spline along the meeting sides of the two boards (see sketch). Cut the grooves slightly deeper than the spline width and in the exact center of the board thickness. The best way to cut such grooves is with a router or a table saw.

Then assemble with glue and clamps.

Another possibility for joining several boards involves the use of dowels. To make this joint, first

make holes in the boards. You can either use a doweling jig or a drill. If you use a drill, first drive

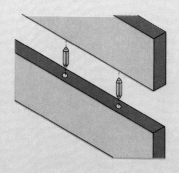

brads (small finishing nails) into one board and press them against the second board to leave marks for drilling. Make the dowel holes slightly deeper than the dowels. Score the dowels, apply glue, join the two boards together, and clamp with pipe or bar clamps until the glue sets (allow plenty of time).

If you'll be drilling many dowel holes, you may want to use a wood or metal template to ensure accurate spacing.

Box Joints

One joint is so common in the construction of boxes — and drawers — it's called a *box joint,* or a *finger joint because its parts* look like the outstretched fingers of a hand (see sketch). Note that one of the mating pieces must have two end fingers, or one more

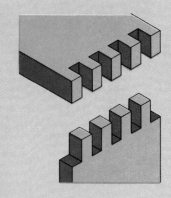

finger than the piece it is to engage. You can make this joint by hand with a backsaw and a small, sharp chisel. However, it is much easier, quicker, and more accurate to make it on a table saw. Use a dado blade set to the desired width and proper depth of the fingers and mark off the waste area so there will be no mistake as to what you want to cut away.

THE HARDWARE YOU'LL NEED

For any sort of fastening work, you will need nails, screws, and bolts, as well as glues and cements.

Nails, Screws, and Bolts

These most common of all fastening materials are available in diverse widths and lengths, and in steel, brass, aluminum, copper, and even stainless steel.

Nails. Nails are sold by the penny—which has nothing to do with their cost. The "penny," (abbreviated *d*) refers to the size. The chart shows a box nail marked in the penny size designations as well as actual lengths in inches.

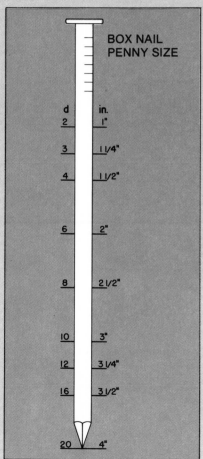

BOX NAIL
PENNY SIZE

d	in.
2	1"
3	1 1/4"
4	1 1/2"
6	2"
8	2 1/2"
10	3"
12	3 1/4"
16	3 1/2"
20	4"

Use common nails for general-purpose work; finish and casing nails for trim or cabinetwork; and brads for attaching molding to walls and furniture.

COMMON SCREWS

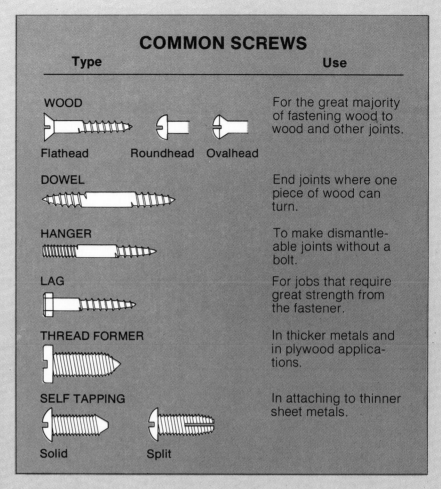

Type	Use
WOOD — Flathead, Roundhead, Ovalhead	For the great majority of fastening wood to wood and other joints.
DOWEL	End joints where one piece of wood can turn.
HANGER	To make dismantle-able joints without a bolt.
LAG	For jobs that require great strength from the fastener.
THREAD FORMER	In thicker metals and in plywood applications.
SELF TAPPING — Solid, Split	In attaching to thinner sheet metals.

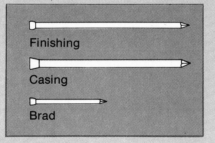

Finishing

Casing

Brad

Screws. Screws are sold by length and diameter. The diameter is indicated by a number, from 1 to 16. The thicker the screw shank, the larger the number. The drawing shows some of the most popular types of screws.

Always drill a pilot hole when inserting a screw into hardwood. And always drill a clearance hole in the leading piece of wood when screwing two pieces of wood together. Without a clearance hole, the leading piece tends to "hang up," preventing a tight fit between the two.

Bolts. You can also fasten wood together with bolts, but only if there is access to the back for the required washer and nut. A bolted joint is stronger than a screwed joint, as the bolt diameter is generally thicker than the comparable screw, and also because the wrench used to tighten the nut can apply much more force than a screwdriver in a screw slot.

Glues and Cements

While not "hardware" as such, glue is an important adjunct to any fastening job. The so-called white glue is excellent for use with wood, and only moderate clamping pressure is required. When dry, it is crystal clear. However, it's not waterproof so don't use it for work subject to excessive dampness—and of course, never for outdoor use. Use the two-tube epoxy "glue" for joints that must be waterproof.

Plastic resin glue, a powder that you mix with water to a creamy consistency, is highly water resistant.

Contact cement provides an excellent bond between wood and wood, and wood and plastic. When working with contact cement, remember that it dries instantly, so position your surfaces

COMMON BOLTS

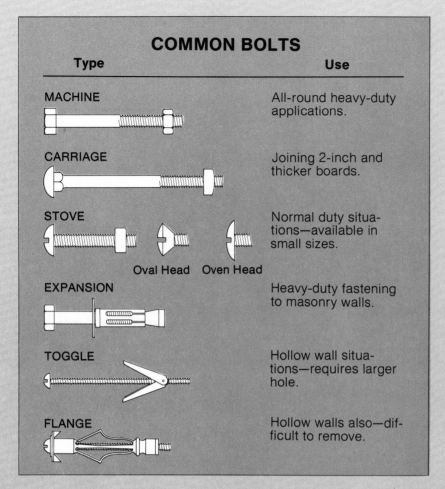

Type	Use
MACHINE	All-round heavy-duty applications.
CARRIAGE	Joining 2-inch and thicker boards.
STOVE Oval Head Oven Head	Normal duty situations—available in small sizes.
EXPANSION	Heavy-duty fastening to masonry walls.
TOGGLE	Hollow wall situations—requires larger hole.
FLANGE	Hollow walls also—difficult to remove.

When to Use What Glue

Type	Use
White glue (No mixing)	Paper, cloth, wood
Epoxy (requires mixing)	Wood, metal, stone (waterproof)
Plastic resin (requires mixing)	Wood to wood (water resistant)
Contact cement (no mixing)	Wood to wood or plastic (waterproof)
Waterproof glue (requires mixing)	Wood to wood (waterproof)

together exactly as you want them. You won't get a second chance.

True waterproof glue comes in two containers; one holds a liquid resin, the other a powder catalyst. When dry, this glue is absolutely waterproof and can be safely used for garden equipment and all outdoor projects and furniture.

Glides and Casters

The intended use determines whether a piece of furniture needs a caster or a glide. If you don't plan to move it frequently, use a glide; otherwise, a caster is the best choice.

Glides come in many sizes, determined by the glide area touching the floor, and with steel or plastic bottoms. The simple nail-on glides aren't height adjustable but you can adjust screw glides by screwing the glide in or out to prevent wobbling if the floor is uneven, or if by some chance, the project does not have an even base.

Casters are made in two styles—stem type (only the stem type is adjustable) and plate type (at left in sketch). The stem type requires a hole to be drilled into the leg or base of the cabinet or furniture. This hole accepts a sleeve that in turn accepts the stem of the caster.

The plate type caster is merely screwed to the bottom by four screws that pass through holes in the plate. They are not height adjustable unless, of course, you use shims.

All casters use ball bearings as part of the plate assembly to facilitate swiveling. For extra-heavy usages, purchase casters with ball-bearing wheels as well.

The wheels on casters are of two types—plastic or rubber. Use casters with plastic wheels if the project is to be rolled on a soft surface such as a rug; rubber wheeled casters are best on hard concrete, vinyl, or hardwood. It's a good idea to use graphite to lubricate the wheels and their bearings, as oil tends to pick up dust and dirt.

To prevent a caster-equipped unit from rolling, get locking casters. A small lever on the outside of the wheel locks a "brake." Brakes on only two of the four casters on a unit are sufficient.

Miscellaneous Hardware

There are many types of hardware that can come in handy when you're constructing storage bins, cabinets, chests, shelves, and other projects.

Following are some you may need from time to time: corrugated fasteners connect two boards or mend splits in wood; metal angles reinforce corners; flat and T plates also reinforce work; masonry nails secure work to concrete or brick walls; steel plates with a threaded center are used for attaching legs to cabinets; screw eyes and cup hooks allow for hanging items inside storage units; and lag screw plugs made of lead or plastic secure furring strips or shelf brackets to masonry walls.

You'll be wise to stock your workshop with most of these items in a couple of sizes. That way, you won't have to make a special trip when they're needed.

HOW TO APPLY PLASTIC LAMINATE

High-pressure laminate, or simply "laminate," is resin-coated paper that has been laminated under high heat and pressure. The result: a versatile rigid sheet that's excellent for covering kitchen and bathroom countertops, furniture, and cabinets.

A Few Words About Laminate

Laminates are available in an extensive range of patterns and colors; some of them are color-keyed to match kitchen appliances and bathroom fixtures. You may have to special-order the laminate you want, but delivery usually doesn't take long.

Standard laminate sheets are ⅟₃₂ and ⅟₁₆ inch thick and measure from 2×5 to 5×12 feet. Buy the thin sheets for vertical applications and the thicker ones for horizontal uses.

You can use regular hand and power tools to cut, drill, trim, and form laminate. However, it's not a bad idea to buy an inexpensive notched trowel to use for spreading on the contact cement. This trowel creates a thin, uniform bed of adhesive that ensures good adhesion. An old paintbrush will work, too.

Taking care of laminated surfaces—whether they're in the kitchen, bathroom, or elsewhere—is no chore. Common sense and a modicum of caution will keep plastic laminate looking good for years. Just remember that although laminate is tough, it can be damaged by high heat and some household cleaners that contain abrasives, peroxide, or chlorine.

Step-By-Step How-To

Plastic laminate is one of those materials that become more intriguing the more you work with them. With practice, you can learn to lay it down as well as many pros. Regardless of the project, the techniques remain much the same. So let's discuss a project you may encounter—re-covering a countertop.

First remove the old covering down to the base material. If the base material is badly damaged, tear it out and replace it with plywood. If the surface is in decent shape, sand it smooth or top it with tempered hardboard.

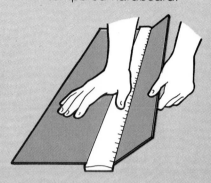

To cut laminate, score its face with a carbide-tipped blade, then hold one side flat, grasp the other, and snap it up.

Or saw the laminate with a fine-tooth back or circular saw. On a table saw, cut with the good face up; with a portable, face-down.

Protect the laminate's edges and corners, because they can chip. And don't bend the material too much; it could snap.

To bond laminate, carefully cut and pre-fit, then apply special

contact cement to the back of the laminate and the bonding surface. Apply all edge pieces first. When the cement is dry to the touch, cover the surface with

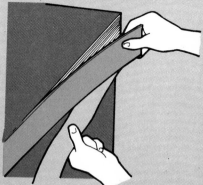

brown wrapping paper, position the laminate, and pull out the sheet. The cement bonds *on contact,* so work carefully. If you do make a mistake, as a last resort fill an oil can with lacquer thinner and squirt it under a corner, peeling the laminate back as you go. Let the thinner dry, and reposition the laminate. You will not need more cement.

When all the laminate is down, roll all surfaces with a rolling pin. Tap along the edges with a hammer and wooden block.

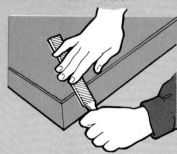

The difference between a professional-looking job and a slapdash job will be in how you trim the edges. You can smooth them by hand with a file, but a router with a laminate bit will save lots of time. File as shown after routing to remove sharp edges.

HOW TO INSTALL CABINET DOORS

Except for shelves, tables, and chairs, nearly every piece of furniture you build will have some sort of door. All doors require hinges or tracks, and handles for opening and closing. Here are the basics.

Construction Pointers

To prevent warping, cabinet doors should be at least ½ inch thick. However, you can use a ¼-inch panel, providing you frame it with ½-inch wood, somewhat like a picture frame.

If you plan to laminate a door panel with plastic, use the thin grade laminate especially made for vertical surfaces. The heavy grade, made for countertops, may cause the cabinet to warp.

Sliding Doors

Sliding doors are easier to fit and install than swinging doors, and, as a rule, are of much lighter stock than conventional doors. Track for sliding doors can be aluminum or plastic (left sketch), or it can consist of grooves cut into the top and bottom of the framework (right sketch).

Of course, you must cut these grooves before assembly. Make the upper grooves about twice as deep as the bottom ones so you can lift up, then lower the door into place. The doors should be flush with the bottom shelf surface when it's touching the top of the upper groove.

To ease sliding, apply wax or a silicone spray to the grooves. If you're planning to use handles, recess them into the door so there will be no interference when the doors bypass each other.

Hinged Doors

Flush-type hinged doors that recess within the framing require clearance all around to prevent binding. To install a flush-type door, make a dry fit, and if the door fits, insert small wedges at all sides to hold it in place and ensure clearance until the hinges have been completely installed.

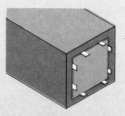

Then place the hinge against the door—if it's an exterior mounting—and mark the hinge holes with an awl. Drill pilot holes and install the hinges. Use this same procedure if you have an interior mounting job.

With hinges that are partly concealed—half on the inside of the door and half on the frame—mount the hinges on the door first, set the door in place, and mark the location of the hinge on the frame or door jamb. This method is much easier than trying to fit an already-mounted hinge to the blind or interior part of the door.

Types of hinges. There are literally dozens of types of hinges to choose from. Following are a few of the more common varieties.

As a general rule, you should mortise hinges into cabinets so they are flush with the work. However, always surface mount decorative hinges, such as colonial, rustic, and ornamental hinges.

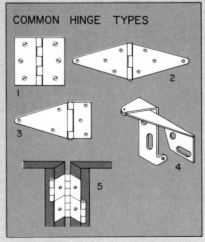

COMMON HINGE TYPES

(1) *Butt hinges* are the type you're probably most familiar with. Use them for either right- or left-hand doors. The larger sizes have removable pins to facilitate taking off the door; the smaller sizes don't. For long cabinet doors or lids. use a piano hinge (a long butt hinge) rather than several smaller ones. (2, 3) The *strap hinge* and the *T hinge* are used for extra-heavy doors. There's no need to mortise these hinges, as they are strictly functional.

(4) *Pivot hinges,* also called knife hinges, are available in different shapes and are especially good for use on ¾-inch plywood doors. All shapes present a very unobtrusive appearance.

(5) *Double-acting hinges* allow a door to be swung from either direction.

Self-closing hinges operate by means of a spring concealed within the barrel of the hinge. Another type, used on kitchen cabinets, has no spring, yet closes the door with a positive snapping action. Its secret is a square shoulder next to the pin.

Special-purpose hinges are available with offset leaves (so the door will overlap the framing); hinges with knuckles (for quick door removal); ball-bearing hinges lubricated for life (for extra-heavy doors); hinges that will automatically raise a door when it is opened (so that it will clear a carpet on the far side of the door); burglar-resistant hinges (with pins that can't be removed when they're on the outside); and hinges that allow a door to be swung back far enough so that the full width of the doorway can be utilized.

Door catches and handles. In addition to hinges, you will need hardware to keep the doors closed and to lock them. For cabinet work, your best hardware bets are spring-loaded or magnetic catches.

Spring-loaded catches come with single and double rollers and are ideal for lipped doors, flush doors, double doors, and shelves. These catches are adjustable.

Install magnetic catches so there is physical contact between the magnet in the frame and the "keeper" on the door.

A handle of some type is required for all drawers and doors. Handles can be surface-mounted or recessed flush with the drawer or door. Sliding doors always use recessed handles so the doors can bypass each other.

FINISHING TECHNIQUES

Finishing is your final job before you can step back and admire your work. Before starting, make sure that all nails are flush or countersunk and filled, all flathead screws are flush with the surface, all cracks are filled, and all surfaces are sanded and cleaned.

Hardboard and Chip Board

If the unit you have built is made of hardboard, about the only finish you can apply to it is paint. No preparation is needed except to remove any oil or dirt. Inasmuch as hardboard is brown—the tempered type is a darker brown—you'll need to apply at least two coats of paint if you want the final finish to be a light color.

Hardboard will accept latex or alkyd paints equally well. Between coats, let dry overnight and then sand lightly.

You also can paint chip board, flake board, and particle board, but because of their slightly rougher texture you should apply a "filler" coat of shellac first, then proceed with painting.

Plywood

Because of its comparatively low cost, fir plywood is used extensively for building projects. However, the hard and soft growth patterns in the wood will show through unless a sealer is used before painting or finishing with varnish or lacquer.

After sealing, sand lightly and finish with at least two coats of paint, varnish, or lacquer. The final step for varnish or lacquer work consists of an application of paste wax applied with fine steel wool and polishing with terry cloth or any other coarse-textured cloth.

Plywood has a pronounced end grain due to its layered construction. If your project will be on display, it's best to hide the end grain, and there are several ways to do this.

A mitered joint is the obvious solution, as then the end grain is hidden within the joint. Another solution is wood veneer tape (see sketch). This tape comes in rolls and is really walnut, oak, mahogany, or a similar wood in a very thin strip about ¾ inch wide. Either glue it or use contact cement, applying the cement to the tape and to the plywood edges. When the cement has lost its gloss, carefully align the tape and press over the plywood edge.

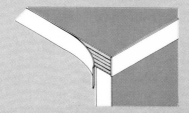

You also can use molding to cover the edges. It has the additional advantage of making a decorative edge requiring no further treatment.

Metal molding is another option, especially useful for edges which are subject to wear and abuse.

A rabbet joint will also hide end grain. Make the rabbet deep enough so that only the last ply is uncut.

Other Woods

If your project is constructed of a fine wood, a more elaborate finishing technique is needed.

Sanding. You can do this by hand or with a power sander. A power belt sander is fine for initial sanding, but always do the final sanding with an orbital or straight line finishing sander—or with fine sandpaper.

Filling and staining. Open grain woods such as oak, chestnut, walnut, ash, and mahogany require a filler to close their pores. Apply the filler with a brush or rag, wiping across the grain. After 10 or 15 minutes, remove the excess filler with a coarse cloth.

If a stain is called for, let the wood dry for 24 hours before application. A stain applied over a filler that has not dried will show up as a "hot" spot.

Sealing. A sealer, as its name implies, is used to seal the stains and filler from the subsequent finishing coats.

One of the best sealers is shellac. One advantage of using shellac is that it prevents the stain from bleeding. Thin the shellac with alcohol to the consistency of light cream; as it comes in the can, it's much too thick for use as a sealer. You can also use ready-mixed stains combined with a sealer.

Finishes. *Varnish,* the traditional finish for wood, is available in many types and colors.

To prepare a piece for varnish, sand it lightly, wipe off the dust with a turpentine-dampened rag, and apply the varnish with long, flowing strokes. Do not brush out the varnish as you would paint. And don't use varnish during humid weather. To make sure the varnish will flow evenly, place the can in warm water.

Varnish requires at least two coats, with a minimum of 24 hours drying time. Sand lightly between coats. After the second or third coat has dried for at least a week, rub down with steel wool and paste wax. Polish with a rough cloth.

Shellac, too, will yield super results. It's fairly easy to work with and it dries dust free in a half-hour. You can apply the second coat within two hours. Sanding is not required between coats, as the second coat tends to partially dissolve and melt into the first one.

One disadvantage of shellac is that it shows a ring if a liquor-stained glass is placed on a shellac-finished surface. Also, shellac sometimes tends to crack if exposed to dampness.

Polyurethane is a tough synthetic varnish that resists abrasion, alcohol, and fruit stains. It's great for floors, furniture, walls, and woodwork. To apply polyurethane the surface must be clean, dry, and free of grease, oil, and wax. Don't apply a polyurethane finish over previously shellacked or lacquered surfaces. Allow at least 12 hours drying time for each coat, and clean your brushes with mineral spirits or turpentine.

Lacquer is a fast-drying finish you can apply by spray or brush. For spraying, thin lacquer only with lacquer thinner. *Never use turpentine or mineral spirits.*

To brush lacquer, always use a brush that has *never* been used to apply paint.

And never apply lacquer over a painted surface, as the lacquer will lift the paint. As with shellac, sanding between coats is not necessary.

HOW TO UPHOLSTER CUSHIONS

By now, you've probably stepped back to admire your new furniture project. But if you have any doubts about its comfort, here's your chance to make your project as cozy as it is attractive.

Pieced Cushion Covers

For a pieced cushion cover, you'll need a top, bottom, and a strip for the sides (called the boxing strip). For a square or rectangular cushion, begin by measuring the width and length of the top and bottom (see the sketch below). For a round cushion, measure the diameter and draw a circle to size to serve as a pattern.

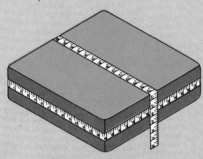

To determine the size of the boxing strip, measure the thickness of the cushion and its outside perimeter, then cut the fabric to these measurements. Don't add seam allowances. As you stitch the cover, use ½-inch seam allowances to make the cover 1 inch smaller than your cushion. This ensures a snug, neat fit.

For a cushion with a zipper, cut a length of boxing strip 1 inch longer than the zipper. Cut this piece lengthwise through center. With the right sides of the fabric facing, machine-baste one long edge together in a 1-inch seam. Press the seam open and insert the zipper.

To determine the length of the remainder of the boxing strip, subtract the length of the zipper from the total length of the boxing strip. Add 1 inch to this measurement. Using ½-inch seam

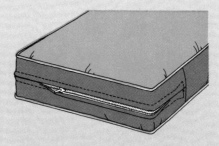

allowance, join the ends of this boxing strip to the zipper section (see sketch). Open the zipper. Stitch the boxing strip to the top and bottom pieces. Trim seams. Turn the cover right side out, press, and insert cushion.

To cover a round bolster, cut fabric circles for the end pieces the same size as the diameter of the bolster. Measure the length and circumference of bolster. Cut a fabric rectangle this size. Stitch the circles to the rectangle. To make a removable cover, you'll need to insert a zippered strip in part of rectangle.

One-Piece Covers

This technique requires less time and fabric than making a pieced cover. To determine the amount of fabric needed, wrap the tape snugly around form to determine width (A) and height (B). Measure the depth (C) as shown. Use the full amount of measurement B, plus a 1-inch seam allowance for the length of the rectangle, and ½ of measurement A, plus a 1-inch seam allowance, for the width of the rectangle.

Mark off a rectangle of this size on wrong side of fabric. Cut out the piece. You'll need a zipper that is 2 inches shorter than cushion's width.

To assemble the cover, fold the rectangle, right sides together, in half crosswise. If your cover has a zipper, center the zipper on the ½-inch end seam and mark the zipper opening. Using ½-inch seam allowances, stitch to these points from the

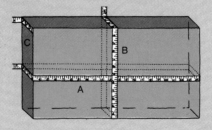

edges and reinforce with backstitching. Machine-baste the opening and press open.

Center the zipper over the basted seam and baste. Turn the cover to the right side and stitch the zipper in place. Remove the basting and open the zipper. Turn cover to the wrong side again, and stitch the side seams with ½-inch seam allowances.

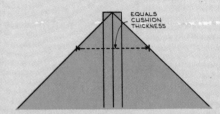

To shape each corner, fold one corner with the side seam centered, as shown. Across the point, draw a line that equals the depth of the cushion form (measurement C). Cut off the corner ½ inch above this line and stitch on the line; backstitch to reinforce.

Cushion Forms

Ready-made forms are available in a variety of shapes and sizes. Box-edge cushions, which can be square, rectangular, or round, give the best results for most furniture projects.

Most box-edge forms are slabs of polyurethane foam available in several thicknesses. You can cut these easily to fit your project by using a razor blade and a straightedge. And, for a softer cushion, stack two or more thicknesses of thinner foam together.

Bolster forms usually are molded and come in various sizes, lengths, and shapes (triangular, tubular, wedge, or five-sided).

However, you can construct almost any shape cushion if you make your own form. First decide on the size. Use muslin as the cover material, following directions given above for upholstering cushions. Assemble and stuff muslin forms.

For filling, use shredded foam, polyester fiber fill, or kapok. Or use cotton batting that's sold in a roll about 16 inches wide and 1 inch thick. You can cut and stack it, or roll it to fit your dimensions. (Cotton batting is also good for use as a filler to adjust foam forms to a particular size).